DMX For Mobile DJs – The Essential Guide for Beginners

By Jordan Nelson

Text Copyright © 2018 Jordan A Nelson

All Rights Reserved

CONTENTS

	Preface	1
Chapter 1	A Little History	Pg 4
Chapter 2	How Does DMX 512 Work?	Pg 8
Chapter 3	Purchasing Fixtures With DMX	Pg 18
Chapter 4	Choosing a DMX Controller	Pg 31
Chapter 5	Some Important Terms	Pg 42
Chapter 6	Setting Up Your DMX Lights For Programming	Pg 44
Chapter 7	Programming on a Hardware Controller	Pg 51
Chapter 8	Programming on a Software Lighting Program	Pg 65
Chapter 9	Scene Builder and The Generator	Pg 82
Chapter 10	Putting it All Together; Tips, Tricks, and Workflow	Pg 90
	Conclusion	

Preface

When I released my first book, "The Essential Guide to Building a Mobile DJ Light Show," my goal was simple; to provide a condensed, easy-to-understand guide that would help anyone new to DJing (or new to DJ lighting) to quickly learn the fundamentals of creating a light show that would set them apart from the sea of other DJs in their market. This  book has been greatly received, and I've tried my best to collect as much feedback as I can from DJs on the internet, here in my hometown, and across the country about how I can improve it in future editions. One of the biggest requests that I have received, however, is for an entirely new book. A book that deals exclusively with DMX lighting and how to incorporate it into your mobile DJ business. The second edition of that book is what you now hold in your hands (or stare at on your computer screen if you bought the eBook).

I gained my passion for lighting when I was 16 and first made my way into the mobile DJ world. In the beginning, simply owning a plethora of lighting fixtures was enough to satisfy my craving, and I would happily mount them all on a stand at each event and beam from ear to ear at my production prowess. Eventually, I realized the same thing that many that have come before me realize with time - that sound-activated or automatic lighting can become boring quickly. And I mean REALLY quickly. I can't put too much of the blame on the manufacturers or the programmers, as you can really only put so many macro options into a certain light before the time and equipment investment exceeds the benefit. I knew that I wanted to do more with my lighting fixtures because I knew that if I was getting bored with my lighting then my clients would also get bored.

And so, I began my intense, OCD, over-the-top search for knowledge on the subject of DMX lighting. I read every book that I could get my hands on, watched every video on YouTube, and visited every forum I could to glean ideas from the experiences of others. Some had phenomenal ideas; others, not so much. I couldn't tell you (or even begin to count) the number of hours I spent watching tutorial after tutorial, gig log after gig log. Now, 7 years later, my brain is full to the brim with everything DMX and I've decided to create a resource that I wish I would have had 7 years ago when I started.

This book is broad and covers topics you may have not even considered or contemplated. Yet it is also succinct; it includes clear descriptions and step-by-step guidance. It does not delve into the deep end of lighting design - for a mobile DJ that comes through experience and experimentation. Instead, it provides you, the mobile DJ, with a base of knowledge that will allow you to confidently purchase lights and a controller and go on to create visually impressive lighting displays through DMX programming.

In this second edition, I have added new, helpful tips below certain sections that summarize important information for quick reference when you need a refresher. **Look for them in bold, *italicized font.*** I've also added practical examples and scenarios to help further illustrate certain points. *Look for them in italicized font.*

Hopefully, you will use those skills to expand, grow, and strengthen your business. While I can't promise that you'll become an expert after only one read, I can assure you that you will increase your professionalism with practice and an open mind and that DMX will be a lot less confusing than before you picked up this book. Thank you for purchasing this book, it means a lot to me! I hope you enjoy it, and here is to the start of your DMX journey!

-Jordan

Chapter 1 – A Little History

It's a great time to be alive

Before the invention of DMX, lighting was analog. Prior to its creation, lighting fixtures were controlled with individual buttons or switches, or even levers on the fixtures themselves. The only parameters that could be adjusted on these lights were on, off, and dimming. During a show, it could take several people to move these levers and someone else to coordinate them. As you can imagine, it was a very labor-intensive process! As time went on, control wires were run from each fixture to a control console, where individual lights could be controlled from a central area. The only problem? If you had 300 fixtures, you had to have 300 control wires. Can you imagine having to wire that setup? The next step in this evolution was the creation of a single control cable along which multiple signals could be transmitted, reducing the volume of wires running between the control area and the stage. Each manufacturer developed their own protocol, but with time DMX512 was adopted as the standard method of linking controllers to intelligent lighting, fog machines, and other special effects.

An old lighting dimmer board.

DMX512 stands for "Digital MultipleX 512" and was developed by the Engineering Commission of USITT (the United States Institute for Theatre Technology) in 1986, and then revised in 1990. Since its creation, its uses have expanded from its initial installations in theaters to include concert lighting, electronic billboards, and even Christmas lighting! For many years, it was mostly found in theaters and professional concert touring applications and not in the mobile DJ and consumer markets. Before DMX control was being placed into more budget-friendly lighting options for mobile entertainers, most DJs or small bands used "chasers" or dimmer packs to add variety and movement to their light shows. These control systems were akin to the automatic programs present in current lighting, but even more monotonous. While they were definitely a step in the right direction, the amount of control over the look and feel of each event was pretty limited. Nowadays, almost every fixture made by a major lighting manufacturer includes DMX control options. That is why the subtitle of this chapter states that it's a great time to be alive; there have never been more lighting options available for mobile DJs!

So why does all this history matter to the mobile DJ? I can sum it up in just a few words - it makes your life much simpler! Now, you are able to control as many lighting fixtures as you want or need with a single cable from your lights to your controller. And we can't forget the controller; most are small, lightweight, and easy to transport which makes for a simple load in and load out. If you opt to go the software route, your controller can be as slim as a laptop! Having DMX means that each of your lights, no matter which manufacturer they are from, will work together. As mobile DJs, being able to quickly and efficiently move and setup gear is of high importance.

One of the unfortunate realities of the mobile DJ industry is that a vast majority of DJs, both new and veteran, are lazy. Too many seek out the easiest and most hassle-free solutions to their problems, and this mindset is easily visible in the sloppy lighting videos that dot the Facebook profiles and groups of your local event market. For a DJ simply seeking to make a buck or get by on the lowest tier of the event ladder, perhaps this is fine. You, however, are looking to rise above the sea of mediocrity. The fact that you have picked up this book is evidence of your desire to stand out and separate yourself from the beginners and the lazy alike. When utilized correctly and effectively, DMX lighting can be a game-changing addition to your services. It can transform your "light show" into a production, and can even alter the entire image of your company.

Adding DMX lighting to your roster of services can be daunting, especially when one considers the additional space, time, and physical effort required to produce quality results. Luckily, we live in the "golden age" of DJ lighting and, with a little research and practice, it is simple and painless to add professional lighting services to your business.

Chapter 2 – How Does DMX 512 Work?

It's not rocket science, I promise

It's no wonder that so many DJs are apprehensive to learn DMX - even its name is imposing! Plenty of DJs will say "It's so confusing! I've watched a million videos and it just isn't clicking!" or "I took one look at a DMX controller and felt overwhelmed." If you've had those thoughts, then you are in the right place. We are going to take a slow, step-by-step approach to understand how DMX works, from the controller all the way up to the last light of your light show.

So, let's break down the name DMX-512: Digital MultipleX (the DMX part of the name) simply means that the signals used to control the lights are sent through a digital communications network. Instead of a dimmer unit physically controlling the amount of electricity sent to a fixture, the controller sends out electronic signals that can be interpreted in the fixture itself and translated to changes in the output of the fixture. Imagine it as one computer (in the controller) telling another computer (in the light itself) how to change the attributes (such as brightness or color) of the light. When you are programming on your DMX controller, you are altering the signal sent out of your controller that travels through the cable and into your lights. Within your lights is a system for reading that signal, understanding it, and translating it into changes you can see.

Quick summary: DMX-512 is a control language that your DJ lights and controller use to talk to one another.

So, what does 512 mean? 512 is the number of control "channels" per data link. In the earliest fixtures, such as basic par cans, each of these channels could control dimmer levels. Imagine it as 512 sliders controlling the brightness of 512 light bulbs (but without 512 wires). If the slider is at the bottom (0), the light will be fully off. With the slider at the top (255, or full), the light is fully on. I have no idea who picked 512 as the magic number of channels that would be in a DMX signal, but 512 seems to be a common number in computers doesn't it? 512 channels of DMX make up one "DMX universe." Over time, the number of features within lights has increased beyond simple brightness control. Now, instead of only being able to control only the brightness of a bulb, you can change color, movement, and even focus!

Most lighting controllers will let you control one DMX universe (512 channels). Some cheaper controllers may only allow control over 128 or 256 channels, while high-end controllers will give you the option to expand your control to multiple DMX universes and well over 512 channels.

Quick summary: There are 512 channels of control in one DMX universe. Each channel controls a specific attribute of your light, such as brightness or color.

Another great attribute of DMX-512 is that DMX is a "bus" system. In essence, this means that you don't have to run separate cables from your DMX controller to every single light you own. Instead, the whole DMX signal is transmitted to the first light and then passes through that light to the next light and so on. This allows for "daisy-chaining" your light fixtures. You run one long cable from your controller to your first light and then smaller cables from each light to the next. Taking a look at your light fixtures you will probably notice the "DMX IN" and "DMX OUT" sockets, and this is exactly what they are used for. Daisy-chaining reduces the number of cables needed to connect your lighting fixtures and saves time.

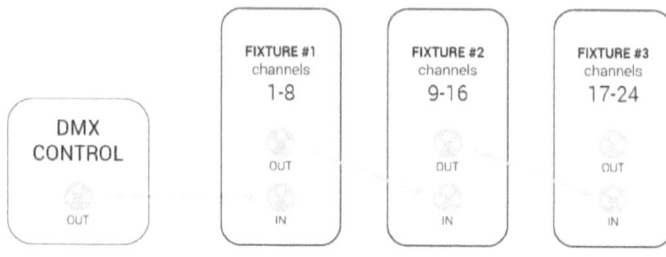

A DMX "daisy chain" linking 3 different lights.

So maybe you have been thinking to yourself "Jordan, if I'm sending out this giant signal of 512 channels that gets passed to all of my lights, how will the light know which part of the signal is meant to control it specifically and not another light?" Ok, maybe it's a stretch to assume you'd think that, but it's the next problem we need to address. Because all of your fixtures are receiving the entire signal, they need to be able to determine which part is for them and which part they can ignore. This is where DMX "addresses" come into play.

Grab your nearest light and turn it around to the back (or side, or wherever the inputs and outputs are). You are going to see one of two things. If your light is older (before the early 2010's) you will notice a row of dip switches. If you've purchased a newer light within the last few years, you'll most likely notice a small LED screen with a few buttons beneath it. Whether you have the dip switches or the screen, these are used to give each fixture an address. An address serves as the starting point for a light to begin reading the DMX signal. We are going to go more in-depth on assigning channels in another chapter, but for now let's imagine it like this: if I set a fixture to channel 16 on that little screen, the light is going to look at the DMX signal you are sending from the controller (I'm sure giving a lot of personality to inanimate objects) and say "Alright, I'm supposed to start listening at channel 16!" The light will only respond to commands from channel 16 through the number of channels that can be controlled in that fixture.

Quick summary: A DMX address is selected within each of your DJ lights and tells the light when to start reading the DMX signal. This allows you to control each light in your show individually if you desire.

The backside of an American DJ Revo 4. Notice the circled area – this screen is where many lighting options can be adjusted, including setting the channel mode and selecting a DMX address.

Did that last part confuse you? Are you feeling overwhelmed? If so, STOP. Take a break. Grab a drink of water and some fresh air if you need. The first time you learn DMX can be confusing and you don't want to get burned out! If you don't quite understand, take a second to start this chapter again and read slowly, making sure you understand each paragraph. Once you feel confident that you understand up to the last paragraph, continue on.

I mentioned 2 paragraphs ago that once a fixture has a DMX address, it will begin reading the DMX signal at that assigned address. So how does it know when to stop reading? This is determined by the light itself. Each light has different attributes. For example, a simple LED par may only let you control its brightness and whether the red, green, or blue diodes are on or off. If you reread that sentence again, you will count 4 parameters of control: brightness, red, green, and blue. This means that if you set that fixture's address to 16, it would know to only read channels 16 through 19 (because it only has 4 parameters of control). Other lights, like a moving head, can have 12 or more channels of control (pan, tilt, zoom, color, strobe, etc.). The more things about your light that you can control, the more channels it will use up.

Some lights that use up a lot of channels will have a setting that can be adjusted on the LED screen to change the channel "mode." If you don't want to deal with the complication of programming 20 channels in a light, you can change the mode which will alter how many channels the fixture is using up. Again, if that did not make sense, reread it slowly. It may seem confusing, but understanding this first chapter is critical to smooth sailing in the rest of the book.

Quick summary: Lights that have a large number of controllable channels (color, brightness, movement, focus, etc.) have an option to reduce the number of channels used by changing the "channel mode." This will reduce the number of attributes you can control but will free up channels in your controller if you have many lights.

Ok, so what have we established? Let's outline the main points:

- DMX-512 stands for Digital MultipleX 512, which means that it is a digital communications network capable of controlling 512 parameters/channels.
- A DMX controller sends out a big package of 512 channels of signal along a single cable to a daisy chain of lighting fixtures.
- Each fixture is assigned an address that lets it know when to start reading the signal.
- Each fixture can be set to a channel mode that changes how many of the channels the fixture is using to be controlled.
- Different lights use different amounts of channels. The more channels a fixture has, the more options you have in creating different looks with color, brightness, movement, and more.

So how does all of this go from tiny bits of electronic data to an epic light show? It happens through the creation of **scenes** and **chases**. I'm going to use photography and videography as analogies to explain these two terms. A **scene** is like a snapshot; it is a "picture" of your light show that can be saved and later recalled during a performance. Let's imagine that you love the song "Fireball" by Pitbull and want to make an awesome red look for your light show to go with the song. You hook all your lights up, jump on your controller, and adjust different channels (brightness, color, etc.) so that your lights are red. You can technically leave all your sliders in those positions, but then you won't be able to program any more colors or variety. If you want to make a new look with a different color, you would have to move all of your sliders and you would lose the red scene you had just created. But what if you like the red look and want to use it again during your event? Having to move the sliders around to program this look multiple times per event (or even every song) would be a HUGE hassle.

Instead, you will save your settings as a **scene** and assign that scene to a button on your controller. Now, you can continue programming other colors and movements, and whenever you want to go back to all red lighting you can simply hit the scene button you programmed earlier and your lights will snap back to those settings! Pretty nifty huh? Scenes save you the trouble of having to manually move sliders and buttons every single time you want to switch up the look of your light show. You can save scenes that have different colors, different levels of brightness, or any other combination of attributes. Then, during an event, you can call up those scenes at will as needed.

Quick summary: Scenes save time by taking "snapshots" of your lighting after you have created different looks. You can recall these scenes rapidly during an event without having to program them again.

A **chase** is like a video; it is a collection of scenes played one after another. Just like going to the movie theater to watch a slideshow might be boring, an event that contains only static scenes might lack the energy and movement desired. Chases are simply the programming of multiple scenes to play back to back with varying speeds. Chases can be built manually on a hardware DMX controller or automatically within a DMX software program. Chases are practically essential if you're using scanners or moving heads and want to create movement that is fluid and smooth. The duration of each "step" or scene in a chase can be adjusted as well as the time it takes to fade or snap between each scene.

Adjusting these variables yields endless combinations.

Quick summary: Chases are multiple scenes strung together and played back. They can be simple (switching between a few colors every few seconds) or complex (many lights moving in defined patterns in unison). They can be created manually or automatically through certain DMX controllers.

When chases and scenes are dynamically used together, you are able to add variety and variability to your light show. From the feedback I have gathered from multiple DJs, it seems the biggest complaint each has about their lighting is the lack of diversity in the programs available to them in sound active or automatic mode. Using a combination of programmed DMX scenes and chases you can create potentially limitless combinations of color, movement, and activity. The best part about the creation of DMX scenes and chases is that YOU are in control of the mood created by the lighting. You can tailor your lighting to your style, your market, or even a specific event.

If you have been hired for a neon 80's themed party, you can program brightly colored scenes into your controller or program. If you are tasked with providing sound and lighting for a jungle-themed corporate event, you can program deep blue and green fades into your lighting. This additional attention to detail stands out to your clients and could potentially influence them to hire you again. With time, lighting may even become a separate service that you are hired for apart from your DJing! DMX lighting control can become a great source of additional income for your entertainment business.

Let me be the first to say congratulations! If you have understood everything up to this point, you have a basic understanding of what DMX is and how it works! See, it wasn't so bad after all. It's probably similar to the first time you gazed upon a large DJ mixer or like the reaction of your friends when you open your mixer case and they behold the vast number of buttons, knobs, and sliders arrayed across its shiny surface. Initially, it looks confusing and messy. But you and I both know that underneath that metal or plastic shell lies a simple purpose for each one of those controls that allow you to create your mix. So it is with lighting.

Chapter 3 - Purchasing Fixtures with DMX

Have the right tools for the job

While purchasing lights was a topic I covered in my previous book quite extensively, I thought it would be valuable to discuss selecting lights that will serve you well in a DMX-oriented role. You may already have lights that are DMX capable and aren't interested in purchasing more lighting at the moment. If so, you can return to this chapter when you are ready to make your next lighting purchases. For the beginner DJ, purchasing lighting will be based more on the need to have fixtures that can serve a wide variety of purposes in order to extract the most "bang for your buck" as you build your business. For the DJ interested in jumping into DMX, your needs may be slightly different. Instead of simply purchasing lights that "look cool" or are portable, a DJ will need to perhaps delve into the manual to examine the DMX attributes of each light, its available channel modes, pixel mapping abilities, and more. In this chapter, I will lay out some guidelines, tips, and suggestions for purchasing lighting that will be capable of handling your needs and wants.

Is the light DMXable?

This one sounds almost too obvious, but it's valid none the less. While most manufacturers are implementing DMX control into an increasing number of lights they offer, there are still fixtures available to purchase that do NOT include DMX control. As you can imagine, these are often the entry-level, budget lighting options that can be had for around the $100 mark. While there is nothing wrong with these lights (many of you may have purchased them when you first got into lighting), they won't serve your purposes if you're now planning to move into the DMX realm. Likewise, if you have purchased a collection of older lights from another DJ or a seller of used DJ gear, they may not have DMX capability. Purchasing used lighting can be a great way to save money, but always make sure that the light you are purchasing is capable of serving its intended purpose.

Example: The American DJ Vertigo Hex LED is one of the most popular styles of DJ light around. You'll see it everywhere from skating rinks to bowling alleys. This classic effect, called a "mushroom," is very eye-catching – however, this popular light does not have DMX control capabilities. While you may enjoy the effect it provides, you would not be able to integrate it into a DMX-controlled show. There are other lights that offer a similar effect to the Vertigo that also have DMX capabilities, and these would be the models you would want to seek out.

How many channels of DMX control are available?

When you first begin to research a light before purchase (you do that right?), you should head straight to the product specifications section of the website or catalog you are planning to purchase it from. If you're researching the light directly on the manufacturer's website, this information is usually available right below the product description. Look for "DMX Channels." Here, you will find 1, 2 or multiple numbers indicating the number of channels (parameters) available for control in that light. Remember channel modes from the last chapter? That is exactly what you are looking at. When the product specs read *"DMX Channels: 7 or 23"* that means that you can select to control 23 channels of the fixture or, for simplified programming, 7.

Description

Chauvet Wash FX2 LED Wash & Effect Light

Wash FX 2 is a multi-purpose effect light with 18 Quad-color (RGB+UV) LEDs that can be used as a regular wash light or eye-candy effect with 6 chasing zones. Powerful UV LEDs create a psychedelic blacklight effect and provide brilliant color-mixing capabilities. Wash FX 2 can be controlled with the IRC6 wireless remote for wireless triggering or in stand-alone, DMX, IR or master/slave modes.

Specifications:

DMX Channels:
- 3, 8 or 28

DMX Connectors:
- 3-pin XLR

Light Source:
- 18 LEDs (quad-color RGB+UV) 6 W, 2 A, 50,000 hours life expectancy

Strobe Rate:
- 0 to 25 Hz

PWM Frequency:
- 78.126Hz

Illuminance:
- 1,795 lux @ 2 m

Power Linking:
- 6 units @ 120 V, 12 units @ 230 V

A snippet of a product listing on a gear website for a Chauvet Wash FX 2. Notice under the specifications that the light has 3, 8, and 28 channel modes.

You may be curious as to exactly what those 23 or 7 channels control, and for that, you'll need to head to the product manual. Nowadays, every major manufacturer releases their product manuals online as a PDF document that can be accessed from any electronic device. A quick Google search should lead you directly to the manual. Crack it open and scan through it until you find the section about DMX programming. Here you will find a table of numbers and descriptions like the one below that describe what each channel of the fixture controls.

DMX Channel Assignments and Values

23CH

Channel	Function	Value	Percent/Setting
1	Master Dimmer	000 – 255	0–100% (only channels 6 to 23)
2	Auto Program	000 – 005	No function
		006 – 020	Pattern 1
		021 – 035	Pattern 2
		036 – 050	Pattern 3
		051 – 065	Pattern 4
		066 – 080	Pattern 5
		081 – 095	Pattern 6
		096 – 110	Pattern 7
		111 – 125	Pattern 8
		126 – 140	Pattern 9
		141 – 155	Pattern 10
		156 – 170	Pattern 11
		171 – 185	Pattern 12
		186 – 200	Pattern 13
		201 – 215	Pattern 14
		216 – 230	Pattern 15
		231 – 245	Pattern 16
		246 – 255	Sound-Active
3	Auto Program Speed	000 – 255	Slow to fast
4	Strobe	000 – 255	0–20 Hz, slow to fast
5	Auto or Sound Program	000 – 010	No function
		011 – 127	Auto
		128 – 255	Sound-Active
6	Red Zone 1	000 – 255	0–100%
7	Green Zone 1	000 – 255	0–100%
8	Blue Zone 1	000 – 255	0–100%
9	Red Zone 2	000 – 255	0–100%
10	Green Zone 2	000 – 255	0–100%
11	Blue Zone 2	000 – 255	0–100%
12	Red Zone 3	000 – 255	0–100%
13	Green Zone 3	000 – 255	0–100%
14	Blue Zone 3	000 – 255	0–100%
15	Red Zone 4	000 – 255	0–100%
16	Green Zone 4	000 – 255	0–100%
17	Blue Zone 4	000 – 255	0–100%
18	Red Zone 5	000 – 255	0–100%
19	Green Zone 5	000 – 255	0–100%
20	Blue Zone5	000 – 255	0–100%

A DMX table from the manual of a Chauvet Wash FX

This DMX table from the manual of a Chauvet Wash FX, for example, has "23CH" at the top left, indicating that it contains values for the light as controlled in 23 channel mode. The left column contains the channel numbers, the second column the specific function of that channel, the third column the values of the channel (remember that DMX channel values range from 0 to 255), and the fourth column contains the specific change associated with each channel value. If a channel controls a color, for example, you will notice that as you move the slider from 0 to 255 that the amount of that color goes from 0% to 100%. In the above table, channel 2 controls the use of built-in programs. Moving slider 2 to a value between 6 and 20 will activate pattern 1, and moving it further will activate other patterns.

Why is this important? Knowing the exact DMX functions of lights that you intend to purchase can help you make sound buying decisions. If you are seeking a certain effect from a light, will you be able to recreate it? Reading through the manual will allow you to gain more insight into whether the fixture will be capable of bringing your vision to life or spend its time sitting in a closet.

Example: Have you ever had a client ask for the "disco ball" or "star ball" effect? With retro-inspired events and parties becoming more and more popular, clients have repeatedly asked me for the effect provided by a disco ball. Many older DJs may still own their mirror balls, but they can be large, unwieldy, and difficult to mount. There are a few new lights available, however, that recreate the effect of slow-moving points of white light with LED technology on a smaller scale. Unfortunately, you will most likely need DMX control to access the ability to control the speed of the movement and the color, and not every "mirror ball" effect light gives you this ability. A quick glance through the manual will let you see if you can control the channels that govern these attributes. Take a look at the following picture which comes from the manual of a mirror ball-style light. If being able to create this look was important to me, I'd want to look for the circled channels.

3Ch

CHANNEL	FUNCTION	VALUE	PERCENT/SETTING		
1	Color Macros and Auto Programs	000 ↔ 005	Blackout		
		Zones	1	2	3
		006 ↔ 013	Red	Red	Red
		014 ↔ 022	Green	Green	Green
		023 ↔ 031	Blue	Blue	Blue
		032 ↔ 040	White	White	White
		041 ↔ 049	Red	Green	Blue
		050 ↔ 058	Red	Green	White
		059 ↔ 067	Red	Blue	White
		068 ↔ 076	Green	Blue	White
		077 ↔ 085	Red Green	Red Blue	Red White
		086 ↔ 094	Green Red	Green Blue	Green White
		095 ↔ 103	Blue Red	Blue Green	Blue White
		104 ↔ 112	White Red	White Green	Blue White
		113 ↔ 121	Red Green	Red Green	Red Green
		122 ↔ 130	Red Blue	Red Blue	Red Blue
		131 ↔ 139	Red White	Red White	Red White
		140 ↔ 148	Green Blue	Green Blue	Green Blue
		149 ↔ 157	Green White	Green White	Green White
		158 ↔ 166	Blue White	Blue White	Blue White
		167 ↔ 175	Red Green Blue	Red Green Blue	Red Green Blue
		176 ↔ 184	Red Green White	Red Green White	Red Green White
		185 ↔ 193	Green Blue White	Green Blue White	Green Blue White
		194 ↔ 202	Red Green Blue	Green Blue White	Red Green White
		203 ↔ 211	RGBW	RGBW	RGBW
		212 ↔ 220	Program 1 / Sound 1		
		221 ↔ 229	Program 2 / Sound 2		
		230 ↔ 238	Program 3 / Sound 3		
		239 ↔ 247	Program 4 / Sound 3		
		248 ↔ 255	Program 5 / Sound 3		
2	Program Speed	000 ↔ 250	Program speed, slow to fast		
		251 ↔ 255	Sound-active mode		
3	Motor Control	000	Stop		
		001 ↔ 127	Rotate left, slow to fast		
		128	Stop		
		129 ↔ 255	Rotate right, slow to fast		

This light allows you to choose a solid white color and a slow-moving rotation.

Is the fixture able to be pixel mapped?

Pixel mapping is one of the most exciting features to become available on newer lights for the mobile DJ. Pixel mapping allows you to gain control over not just the color of the entire light as a whole, but of the individual LEDs or LED segments within the light. This opens up the door to many more customization options in your lighting programming, as you can create scenes and chases with specific light sections on or off. Let me give you an example. The next picture is of the Chauvet WashFX, a light that I studied a LOT before I purchased.

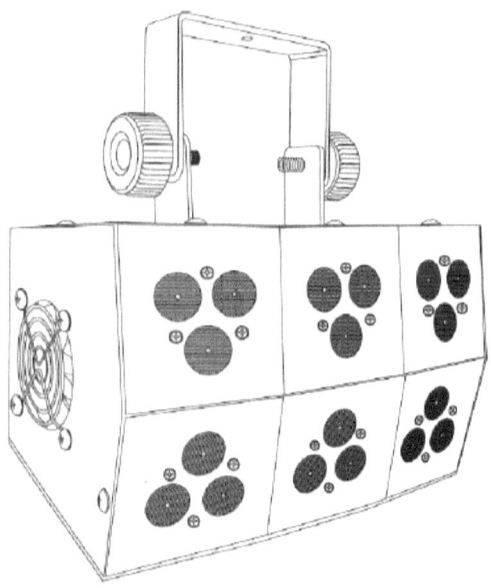

Chauvet WashFX (1st generation)

You'll notice that it has 6 sections of 3 LEDs. The WashFX is pixel-mappable, meaning that I can control each one of those 3 LED sections individually. I am able to adjust the amount of the red, green, and blue color within each segment. This allows me to create chases that move from one section to the other in a wave or create an energetic jumping effect between the various sections. Instead of the entire fixture remaining a solid color, I can create interesting "eye candy" effects.

The ability to pixel map is not essential in a DMX light. In fact, many people feel overwhelmed with the programming of basic lights as it is and don't feel the need to introduce even more complication into the mix. That's fine! This book is meant to be read and re-read throughout your DMX journey, so if in the future you decide to move to pixel mapping-ready lights, you can return to this book to refresh yourself on their function.

Brightness and Color Availability

While both brightness and color availability don't explicitly have an effect on DMX, they are always important when purchasing lighting. Pay attention to the **LUX** values when choosing lights as opposed to the power or watts rating. Many people focus almost exclusively on the number of watts the fixture uses when in reality this statistic does not have as much correlation to the amount of light output it can produce as the lux rating does. When deciding between two or three lights compare the lux ratings and make sure they are stated at the same distance. Lux is measured at a certain distance, usually 1, 2 or 5 meters. It would be useless to say one light is brighter than another if that light is measured at 1 meter and the other at 5.

Specifications:
- DMX connectors: 3-pin & 5-pin
- Source: 7 (5 W) quad-color LEDs, 50,000 hours
- Light source: red (447mA x 2, 10vdc = 0.94W), Green (425mA x 3.25vdc = 1.38W), Blue (459mA x 3.08vdc = 1.41W), Amber (482mA x 2.40vdc = 1.16W)
- Beam angle: 18 degrees
- Field angle: 32.5 degrees
- Illuminance: 434 lux @ 5 m
- Power linking: 22 units @ 120 V / 38 units @ 208 V/ 36 units @ 230 V
- Input voltage: auto-ranging 100-240 VAC 50/60 Hz
- Power and current: 73 W, 0.60 A @ 120 V 60 Hz
- Power and current: 73 W, 0.36 A @ 208 V 50 Hz
- Power and current: 74 W, 0.35 A @ 230 V 50 Hz
- Weight: 5.4 lbs (2.5 kg)
- Size: 9.9 x 9 x 3.9 in (252 x 228 x 100 mm)
- Approvals: CE

The product listing for a Chauvet COLORdash Par-Quad 7. Notice the circled value for illuminance, which indicates the light's brightness

Color availability is my way of saying "Which LED diodes are inside of this light?" RGB diodes are the most common, and the mixing and combining of red, green, and blue gives rise to thousands of other colors. Some lights contain an additional white, amber, and/or UV (ultraviolet) diode to increase the number of colors you are able to produce. I've even seen pink and yellow LEDs in some lights! While these additional diodes are not essential, they can be very important in certain situations. For example, one of the most popular colors in the wedding industry is amber or gold. While a yellow (of sorts) can be achieved on an RGB fixture by mixing both red and green, a rich amber color requires the use of a dedicated amber LED. If you find yourself interested in providing lighting to the wedding industry, purchasing fixtures with amber LEDs would be a great business decision. Once again, examine your goals and plans for your business when purchasing lighting.

Example: When choosing lights, many manufacturers will give you an idea of what color diodes are found in the fixture by how they name the light. If the model name contains the word "quad," it's likely the fixture contains your standard RGB diodes as well as a white or amber diode. If the model name contains "hex," it most likely contains the RGB diodes as well as amber, white, and UV. Be careful, however, as different manufacturers will have different definitions of what a "quad" or "hex" light is. When in doubt, check the listing or manual!

Other Accessories and Essentials

After you have decided on what lights you want to buy (or just using the lights you already have), there are a few more items you will need to start DMX programming. These are a controller (discussed more in the next chapter), cabling, and termination. We are going to save our discussion of lighting controllers for the next chapter because there is a lot to be discussed, so, for now, let's focus on cables and termination.

Cables

One of the most hotly debated topics on DJ forums, YouTube, and the like is whether or not you can use standard XLR microphone cable as DMX cable for lighting. If you take a look at both an XLR cable and a DMX cable you will notice they are strikingly similar.

They have the same connectors on the ends and the same cable between the connectors. So are they different, and if so, how?

To embark on a small side story, let me give you a little background. The DMX-512 standard was originally designed as a 5-pin cable, and if you look at most higher-end lights used in large touring acts you will see a 5-pin in and out connector. This was so that anyone would be able to easily tell a DMX and XLR connector apart, as well as to leave the extra 2 pins open for future expansion. The use of 3-pin connectors is actually prohibited in the DMX-512 standard!

So why do the fixtures you and I use normally contain only a 3-pin DMX connector? When it comes down to it, it's due to manufacturers not really caring too much about the formal rules. They realized that since the 2 extra pins of a DMX connector were not being used (and it doesn't look like they will be used anytime soon) they could design their equipment to use the standard 3-pin XLR connectors to eliminate the need for the unused second pair. While this was a smart move, in my opinion, it has caused no small amount of confusion among many a DMX user. Some argue that there is no difference at all between a microphone cable and a DMX cable and that the two may be used interchangeably without consequence. Others are purists, insisting that only true 5-pin DMX cable will suffice. **It is almost certain that the DJ lights you buy will have 3-pin DMX, so there is no need to purchase 5-pin.**

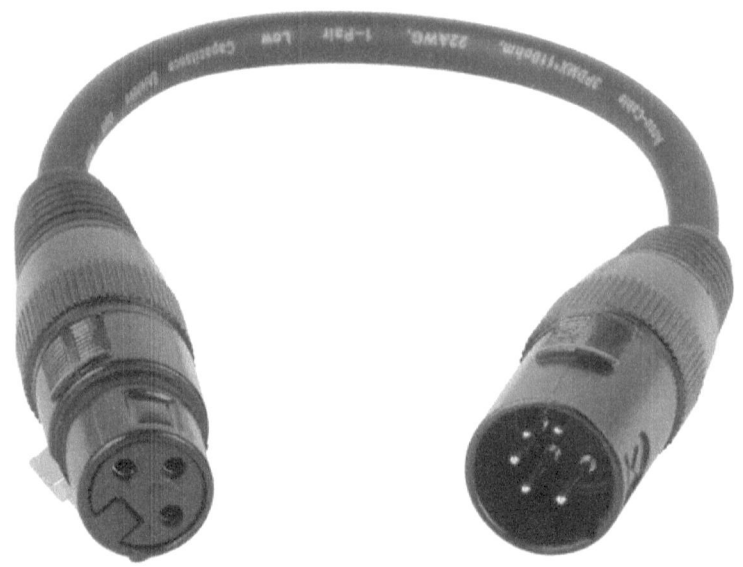

A converter cable with a 3-pin female end and a 5-pin male end. It's unlikely that you will ever need a cable such as this, but it's good to know that they exist.

The fact of the matter is this: while the connectors on XLR microphone cables and DMX cables may be the same, the difference lies in the cable itself. True DMX cable that meets the standard must be 120ohms, and microphone cable comes in lower impedance (usually around 80 ohms). Additionally, DMX cable is better shielded, meaning that the use of XLR microphone cables could allow interference and errors to be introduced into the cable that interrupts the DMX signal.

So why will some DJs swear up one side and down the other that they have used microphone cable for years with their lighting to no ill effect? It's because they have been lucky. In practice, plugging in microphone cable in place of DMX cable for short lighting runs with few fixtures will work, and if you ever find yourself in a rough spot you can substitute the microphone cable in a pinch. The signal will transmit, but you are betting and hoping that the signal will not be disrupted or distorted. So the question I would pose to these DJs is why? Why risk a malfunction in your lighting at an event where a client has paid you hundreds if not thousands of dollars? Why spend hundreds or thousands on a light and then skimp out on a quality cable to connect your lighting? This may just be my opinion, but the difference in cost between microphone and DMX cable is negligible, and to use the improper cable just to save a few bucks is to do a disservice to your company and your clients. Alright, I promise I'll step off my soapbox now!

DMX Terminators

The last component of your DMX lighting system I would like to discuss is the DMX terminator. While not as epic as the name suggests, a DMX terminator plays an important role in the correct function of a lighting chain. The terminator, which looks like the end of a DMX cable (but without the cable), is placed into the "OUT" socket of the last light in your chain. A terminator is especially important when you are using many lights on a single DMX line. It removes a lot of the noise and flickering that can be introduced into the signal over long distances. If the signal travels all the way to the last fixture and there is no terminator inserted, a "shadow" signal can be introduced when the signal bounces back down the line. The successful function of a DMX chain without a terminator is due more to luck than good design. Again, do you want to take that chance at your events? A good terminator costs under $10, so there is really no excuse not to have one.

A DMX terminator.

Chapter 4 - Choosing a DMX Controller

Hardware vs. software

In this chapter, we are going to discuss choosing a lighting controller. I'm not going to sugarcoat it for you; this may be the most difficult decision you are going to make when purchasing lighting equipment. Just like with lighting, there are dozens of options available to you, and settling on one can be a tough decision to make. Luckily, you bought this book! I hope to lay out some of the pros and cons of each system as well as share my own experiences with each so that you can make an informed decision that you feel confident and happy with.

Hardware controllers – The Classic

Hardware controllers are dedicated pieces of equipment that look similar to a DJ mixer. They're full of buttons, knobs, and sliders that give a tactile feel to controlling your lighting. They contain their own electronics and are robust and durable. Some are meant to be set on a tabletop like your DJ mixer and some can be mounted in a rack-style case. The majority of the layouts are similar, so let me walk you through your typical hardware lighting controller.

A basic hardware DMX controller. This unit can be rack mounted or set on a table.

On the left of the controller are buttons representing each of your fixtures. By selecting a button, the controller is now sending a signal to that light and you will see it react to your programming. This prevents all of your lights going crazy at the same time as you program each one individually. To the right of the fixture buttons are the sliders. These control the individual channels of your lights. By sliding each slider up, you are moving that channel's value from 0% to 100%, or from a value of 0 to 255. The controller pictured allows 16 channels per fixture, meaning you can control light fixtures that have up to 16 channel attributes. But you may say, "Why are there only 8 sliders if I can control 16 channels?" Great question! If you look closely to the side of the farthest right slider (don't include the separate set of 2 sliders beneath the LED screen, those have another purpose), you will see a small button. This shift button changes the control of the sliders from channels 1-8 of a fixture to 9-16. Now you can control the other half of your fixture's DMX attributes! However, you won't have to visit this second page of sliders for a lot of fixtures. Many simple DJ lights only use 3, 4, 6, or 8 channels. However, if your fixture has more channels that you would like to use, that shift button allows you to reach them.

Above the sliders, you will see 8 buttons. These are your scene buttons. When you program a light scene you like, you can save it to one of these scene buttons. But don't worry; you can program more than 8 scenes. To the right of that small LED screen at the top of the controller are 2 more buttons with up and down arrows printed next to them. These buttons allow you to change *scene pages*. So if you fill up the 8 scene buttons on page 1 with awesome lighting scenes, simple click down to page 2 and program 8 more scenes! Most controllers allow you to store dozens or even hundreds of pages worth of lighting scenes. You probably won't run out!

As we move to the right side of the controller, we get to more specialized buttons and sliders. The 2 sliders directly below the LED screen control the speed of your chases and the fade time of your scenes (how quickly they transition from one to another). The LED screen shows you what mode your controller is in, what scene and page number you are on, and more. The other buttons are used for programming (entering programming mode, saving scenes, etc.), blacking out the lights, turning on sound active mode, and other specialized functions. Because each controller is different, you'll want to thoroughly read your user's manual to understand your specific controller.

Summary: A hardware DMX controller has different sections that allow you to select a light, change its attributes, save scenes, and program chases. Each controller is slightly different, but using the basic guide above should help you navigate your controller's manual.

What makes a hardware controller good or bad? On the positive side, a hardware controller is self-contained. You don't have to use a computer to control your lighting, which means one less piece of equipment to purchase. Additionally, computers can be finicky and there is always a chance of crashing. It can be argued that a hardware lighting controller is much more reliable and dependable than a computer (although modern computers, if properly maintained, avoid many of those issues). However, hardware controllers do have some major downsides.

One of the biggest drawbacks is the limitations on how many fixtures you can program on a specific controller. Remember how we have 512 channels of control in one DMX signal? With a hardware controller, those 512 channels are often broken up into chunks such as *12 lights with up to 16 channels per fixture.* You'll see wording like that in the specifications section of the manufacturer's website. So, what does it mean? It means that the controller sets channels 1-16 apart for the first light in your DMX chain, 17-33 for the second light, and so on, up to 12 different lights. Even if you are using a simple fixture that only has 4 or 5 channels of control, the next light in the chain will still be programmed 16 channels away. This can lead to a lot of wasted channels and is why the controller is limited to 12 individual lights.

Summary: Hardware controllers can only control a set number of light fixtures, and they also set apart a certain quantity of channels per light they can control. If your light show contains many complicated fixtures, such as moving heads or pixel-mapping lights, you will quickly run out of channels.

The biggest drawback to a hardware controller, however, is the length of time required to program your lights. While practice can improve the speed at which you can program, a large (potentially slow) part of using a hardware controller is having to memorize where each fixture is in the controller, the function of each button, the method for creating and saving each scene, and what each individual slider affects within each light. It requires additional time to switch between pages of scenes and different banks of sliders. When I used a hardware controller, I found myself constantly jumping back to the manual to refresh myself on what each channel did. I'd often have multiple tablets or laptops open simultaneously just so that I could have manuals accessible somewhat quickly. Using a hardware controller isn't a bad option, but it's akin to using a map in the 21st century when GPS is so easily accessible. A map is reliable; the battery won't die and it won't lose service. But it's becoming outdated. GPS systems are more reliable and accurate than ever, and save a lot of time and headache.

*A quick note on larger, higher-end lighting consoles. When speaking about hardware controllers in this chapter, I am referring to the options available commonly to mobile DJs and entertainers that can be purchased for around $80-$300 dollars. Lighting designers and operators in large venues and touring acts use hardware lighting consoles of a different scale than those used by the average mobile DJ. These lighting boards can cost thousands to tens of thousands of dollars and have many more features than the lighting controllers I've discussed. If you have the means and use for one of these boards, your scope of knowledge probably far exceeds the contents of this book.

Software lighting controllers

In the map and GPS analogy I just used, software-based controllers are the GPS. A lighting program can save you immense amounts of time and headache once you learn to master the software. A software-based DMX controller works just like a hardware DMX controller, but on your computer's screen. Looking at a lighting control program, you will see faders and buttons just like you would on a hardware controller. A USB cable connects your computer to a small or box (or dongle) that converts the USB signal into a DMX signal that can be sent to your lights. Your DMX cable plugs into this dongle just like it would plug into a hardware controller.

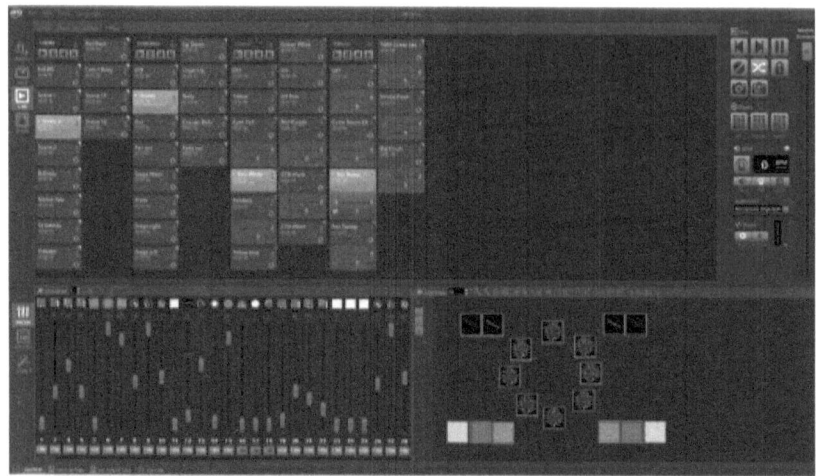

A screenshot of the interface for MyDMX from American DJ.

Using a software-based program has many advantages. With a lighting program, there is no need to be continually consulting your light manuals to determine what each channel or slider does. Major lighting programs used by mobile DJs (MyDMX by American DJ and ShowXpress by Chauvet) have thousands of lighting "profiles" built in. You simply search for and import the lighting profile for the fixture you have and "Ta-Da!" all of the fixture's attributes are added into the software, labeled and ready to go. Some poor sap at the manufacturer has already gone in and labeled each button and slider in the program with its function as it pertains to your specific light. A little phrase or symbol above each slider tells you exactly what the slider does, what value the slider is at, and more.

Once you import your lighting fixtures, you can select your lights with simple drag and drop motions and select pre-programmed colors, chases, and even movements! Creating complex patterns and scenes for beginners is much simpler on a computer than a hardware controller. Instead of manually selecting lights by button, moving multiple sliders to the correct position, and then punching out some button combo to save the scene, you select your fixtures, hit a preset color button, and ctrl+S that sucker to save it! Simple right? And don't even get me started on programming moving heads and scanners with a hardware controller, which involves saving multiple positions as scenes and fiddling with fade and scene times until you get a look you kind-of like. With a lighting program, you can just select your fixtures and select a movement such as a line or a circle. You can even use your mouse to draw out the shape you want your fixtures to trace in the air! Talk about ease of use.

But wait, there's more! With software, you can create and save a limitless number of scenes, chases, and patterns (OK, it may be limited, but I've never reached a limit). What's more, you can add many more fixtures to be controlled by a software controller than a hardware controller. Remember the issue we had with hardware controllers where they dedicated 12 or 16 channels to each fixture even if it didn't need all 16? With a lighting program, you don't have that issue. If your light uses 6 channels you can program it to use channels 1-6, and then program your next light to fit right behind it and start on channel 7! No wasted channels! These are just some of the amazing benefits of using a software-based lighting program.

Summary: A software DJ controller has many advantages, including unlimited scene creation, a much more efficient programming workflow, and premade fixture profiles that save you a lot of time.

But what of the downsides? There is a big glaring one that we already touched on; computers can be finicky. While uncommon if a professional, well-maintained computer is used, a chance always exists of malfunctions, crashing, or running out of battery. Of course, if you are DJing off of a laptop, you're already concerned about that, aren't you? Then there is the obvious barrier of needing a second computer to control your lighting software. I have had great luck using refurbished computers at a much-reduced cost, so there are ways of mitigating the additional expense. Aside from these two downsides, the benefits far outweigh the costs. The reduced programming time (and stress), ease of use, and capability to expand is exponentially higher than when using a hardware controller. If you have not purchased a lighting controller yet, or become frustrated with the one you already own, I HIGHLY recommend going the software route. I've mentioned 2 lighting programs, MyDMX and ShowXpress, which are the programs most commonly used by mobile DJs. Here is what I think of each:

MyDMX

MyDMX is a great starting point for the mobile DJ. The MyDMX buddy dongle with software can be purchased for $99, cheaper than most hardware controllers. It is a scaled down, entry-level version of the full-size MyDMX dongle that can be purchased for $299. It is compatible with both Windows and Mac and allows you control over 256 DMX channels (much more than are needed by the typical mobile DJ). It has an extremely large fixture library; I would be very shocked if you did not find a light you owned in there. If by chance your light isn't in the library, you can reach out to the American DJ and they will create a profile for you. If you don't have the time to wait, you are able to create fixture profiles on your own. It has a very user-friendly graphical interface that makes programming a piece of cake. Almost everything is done in an extremely simple, drag-and-drop effects generator. It's neat because you can actually have visual feedback of exactly what you're creating on your computer screen as you create it.

Whenever anyone asks me what to purchase for their first lighting software I recommend MyDMX. It is by far the most simple, easy to understand lighting software option on the market, backed by fantastic support at ADJ (they aren't paying me for this, I promise). While it is simple to use, it is extremely capable and allows you to create effects that would take hours to create on a hardware controller.

So, what are its cons? Depending on how intricate and custom you want to get with your light shows, you may reach a point where you cannot grow any further with MyDMX. I don't want to get into specifics because the software is constantly being updated, but it does some of the more advanced features that ShowXpress offers for free. Advanced features that are available from ADJ are going to cost you additional upgrade fees (like a 3D visualizer or MIDI control, which is something you will want). Again, for the beginner, these advanced features aren't necessary, but if you decide you want them down the road you are going to have to pay for the upgrades from American DJ.

ShowXpress

After all of the great things I said, you'd expect me to use MyDMX wouldn't you? Well, I did for a long time! It was fantastic software. Eventually, however, I grew to the point where I needed more features, and that's where ShowXpress comes into play. ShowXpress, while requiring a tad more time to master, is very capable of doing some advanced lighting techniques. It allows you to stack scenes, create scene macro buttons, create flash buttons, create light shows along timelines, arrange and customize scene layouts, do pixel mapping, and more. If most of the things I just said are complete jargon don't worry, we will talk about them more.

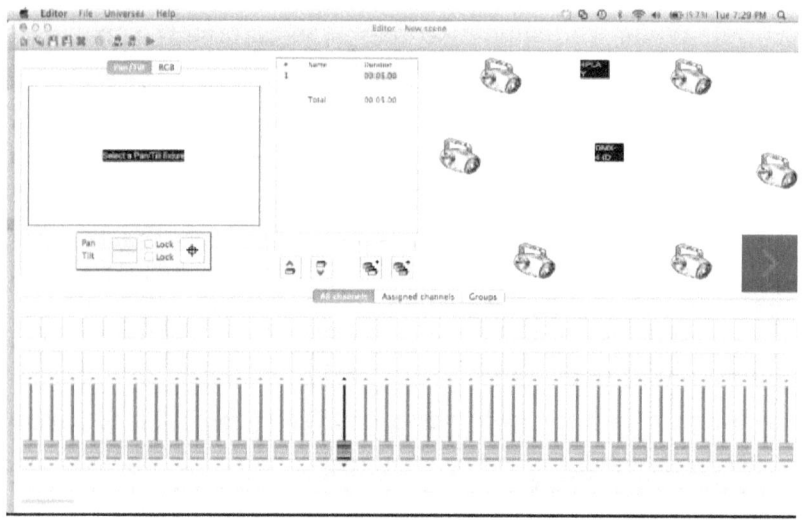

A screenshot of ShowXpress with the channel faders across the bottom and the graphical layout of the lights in the top right.

I would recommend ShowXpress to two types of mobile DJ: Those that are very interested in DMX and plan to grow it to be a significant part of their business and those that are very tech-savvy and can learn computer programs easily. There are plenty of great tutorials online for ShowXpress that can help you learn its functions and features. As a bonus, Chauvet includes the 3D visualizer and the ability to control your lighting with MIDI in the box, meaning no upgrade charge down the road. It allows you to create unlimited scenes and layer as many scenes as you want to live. You can open multiple windows with unlimited pages of scenes in each window. You can trigger scenes by click, keyboard, date and time, automatic, sound, manual BPM, and MIDI. The standard ShowXpress dongle retails for $399, meaning it is going to cost you a bit more initially, but for the advanced user, the features far outweigh the cost in my mind.

Second edition update: Since the first edition of this book was published, American DJ has released the newest version of their software, MyDMX 3.0. The updated program packs in a lot more features that were previously unavailable on version 2.0, including the ability to stack scenes and customize layouts. The playing field has become much more level, and I can now happily say that either ShowXpress or MyDMX will serve you equally. When choosing a program, you will most likely base your decision on personal preference in regards to the graphical interface. Both programs have a free download option that will work without an interface, allowing you to give each a trial run before you purchase.

Chapter 5 – Some Important Terms

Wait, I thought glossaries went at the end?

Before we go on to the chapters where we get down to the nitty-gritty of programming, I thought it might be worth pausing a moment to add some terms to your vocabulary. These are words I'm going to throw around quite a lot in the coming chapters, so familiarizing yourself with them now will save you a few google searches. No, they aren't in any alphabetical order, so just consider them all equally important.

Channel (DMX): A single parameter of a light that can be controlled by a lighting controller, such as color, speed, or strobe.

Channel Mode: The number of DMX channels your light is currently using.

Master Dimmer: A channel found in most DMX lights that controls the overall output of the light, 0 being no light and 255 being fully on, or 100% brightness. This is usually the starting point in programming a fixture.

Scene: A "snapshot" of your lighting after programming has been done. This snapshot can be recalled live during an event.

Auto Programs or Macros: A collection of pre-programmed scenes or chases made by the programmers at the manufacturer. These can be run standalone and/or often accessed through one of the DMX channels.

Values: A sliding scale used in each DMX channel that ranges from 0 to 255. Every channel can be adjusted from 0 to 255. These values can represent percentages (for example, the master dimmer where 255 is equal to 100% output) or steps (for example, any number between 141-155 will activate a certain pattern).

Strobe: A channel that adjusts the flash rate of the fixture.

Gobo: Stencil-like patterns within certain lights (mainly moving heads and scanners) that alter the shape of the beam.

Shutter: A device inside certain lights (mainly moving heads and scanners) the controls light output through physically (or digitally) opening, closing, and strobing the beam.

Pan: The horizontal movement of a moving light.

Tilt: The vertical movement of a moving light.

Chapter 6 - Setting up Your DMX Lights for Programming

Your favorite activity, your wife's least favorite

Alright, it's time to get to it! Assuming you have some lights with DMX capabilities, the correct cabling, and a way to rig your lights up, let's program. But first, you need to make sure you have all the components of your system in order so that you will have flawless functioning and fewer headaches. Whenever I program my lighting, I clear out a nice big space in the living room where I can set up my trussing or lighting T-bars how I would use them at a normal event. If you are not sure how you want to set up your lighting or hang it, check out my previous book for tips on different lighting stands and arranging your lights. Grab your lights, cables, clamps, and truss/stands and get a nice open area where you can set up. This is also a great time to practice your cable management so you have a nice clean presentation at your events.

Start by setting up your trussing or stands and hanging your lights. If you have multiples of fixtures such as pars or moving heads, I suggest hanging them symmetrically, an equal number on each side at an equal distance from each other. Run power to all of your lights and plug them in. Find a comfy chair and table to program from that is off to the side of your stand so that the lighting isn't blinding you constantly while programming. The first step is to power on your controller (or computer and dongle). Plug your DMX cord into the output of the dongle and run the cable to the first light in your light show. It's worth noting here that a longer DMX cable (at least 20 or 25 feet) works best, as it gives you plenty of room to run the cable from your DJ booth or programming area along the ground or stage and up your light stand. Next, grab some smaller DMX cables (3 or 5 feet) and run the next cable from the output of the first light to the input of the second light. Continue this process until you come to the last fixture in the chain. Insert a DMX terminator into this last light.

Remember to make all of your cable connections, set your channels, and adjust the light's positions before you raise your stand. There have been many times I've raised my lights 12 feet up in the air and then realized I forgot to turn something on or plug something in. In fact, at an event, I suggest that raising your lighting stand or truss be one of the last things you do after checking everything is running correctly.

Your lights most likely powered on in automatic mode, so you'll need to change them to DMX mode. The process varies with each fixture but generally involves similar steps. On the screen on the back of your light, push the menu button until the screen displays **"Addr"** or some variation of "address" or "dmx." Your light may only shoe **"d001"** or just the letter "d" followed by another number. Hitting the enter button will take you to a 3-digit number. If this is the first time you have accessed the DMX portion of your light, the default number will be 001.

The menu display is showing the light is in DMX mode (indicated by the letter d) and is set to the DMX starting channel of 10.

If you are programming the first light in your chain, you can leave this fixture as 001 if you would like. Click enter once you have the channel number you would like for the light. Clicking the menu button a few more times will take you from "**Addr**" to "**CHnd**," or some variation of the word channel. This is where you can change the channel mode, and pushing the up or down buttons lets you select from various options such as 7CH or 23CH.

The menu display for a DJ light showing "15CH," which indicates the light is in 15-channel DMX mode.

Refer back to earlier chapters of the book for information on channel modes. Once you have selected the starting channel for the first light and the channel mode you wish to use, you are finished with that light. Make sure to click the menu button a few times until you are back on the screen that says "**Addr**" before moving on. When the light is in DMX mode and is receiving DMX signal through the cable it should be blacked out (i.e. not doing anything).

Now we move on to the second light, and the process here varies slightly depending on whether you are using a software or hardware DMX controller.

- **Hardware Controller**: If you remember back to our discussion on hardware controllers you will recall that every controller allots a specific number of channels to each fixture it can control. For example, *12 fixtures with 16 channels each*. Take a look in your controller manual or the manufacturer's website to find out how many channels each fixture has in your controller. If your controller *does* allot 16 channels to each, you will set your first light to channel 1, and you will set your second light to channel 17. If you have 8 channels per fixture, the first light would be channel 1, the second light channel 9, etc. This process continues on down the chain, up to the number of fixtures your controller can handle. The method of getting to DMX mode, setting the address, and changing the DMX mode are the same as described above; only the actual channel addresses change.
- **Software Controller:** With a software controller, you will set your fixture addressed one after another. If your first

light starts at channel one and uses 6 channels (or stated another way, is in 6-channel mode), you will set your second light to channel 7. If that light uses 10 channels (10 channel-mode), you will set your third light to channel 18. You can continue on down the line in this manner, following the same instructions for setting the address and changing the channel mode as before.

One thing to note is that if you are using multiples of the same light you can assign those lights the same address. For example, if I am using 4 of the same LED par lights for my wash lighting, as long as they are the exact same lighting fixture and you put them in the exact same channel mode you can set them to the same address. This will cause all of them to respond simultaneously to your programming in the exact same way. This saves time whenever you are programming many fixtures, as you don't have to adjust settings for every individual light if they are identical. On the other hand, you may want to have different colors or effects happening simultaneously among your matching lights, which is totally cool too! It's really up to you. If you are limited on the number of lights your controller can handle or you want simpler programming, combining multiples of identical fixtures can make it easy.

Example: Imagine my light show is made up of four of the American DJ Mega Tripar Profile Plus, which we are using in 10 channel mode. I can set each light to its own address – channel 1 for the first, 11 for the second, 21 for the third, and 31 for the fourth. This would allow me to control the brightness, color, and strobe of each light individually with no overlap. Alternatively, I can just set all four of the lights to channel 1. Now, whenever I make adjustments that affect channels 1 through 10 all of my lights will be affected in the same way. This simplifies programming but does reduce flexibility a little.

How you address your lights depends completely on your setup and the outcome you want for your programming. Here are a few sample scenarios and how you might think about addressing your lights.

- **Common wedding setup: 2 totems with an LED uplight inside each one and a moving head on top:** Even with this simple setup of 4 lights you have multiple options for addressing your fixtures. The easiest and most common method would be to assign each moving head the same address (let's say 1) and the pars to the same address (we'll say 9). Now both pars would react the same and both moving heads would react the same. If you wanted to alter the look slightly, you could invert the pan and/or tilt of the second moving head in the fixture settings (look in your product manual for how to invert pan and tilt). This would allow you to program the moving heads simultaneously but the second moving head would move opposite of the first. Another option for addressing your fixtures would be to program each par individually and each moving head individually (a total of 4 fixture addresses). Now the two pars could be programmed independently and each moving head could also be programmed apart from the other.
- **Simple school setup: I-beam truss with 4 pars, 2 moving heads, and a centerpiece light:** Suppose you arrange your fixtures (from the left) *par, moving head, par, centerpiece, par, moving head, par*. Again, you could assign your addresses in groups according to the fixture type. All the

pars could have the same address, the moving heads, and the centerpiece. In essence, this would leave you with just 3 different aspects to program in each scene (the wash, the heads, and the centerpiece). Alternatively, you could program the left two pars to the same address and the right two pars to the same address. Or every other par with the same address. See where I'm going with this?

- **Programming uplighting:** If you do a lot of uplighting and want to control it, the same principles apply. You could have all the fixtures on the same address, changing the color of the whole room together. You could have every other fixture on the same address to create a cool zebra effect. Or you could go in sets of two or three.

Summary: If you have multiples (2, 4, 8, etc.) of the same light you have more channel setting flexibility. Experiment with setting all lights to the same DMX channel, all different channels, or different alternating patterns to give your show variety.

After you have each light hung, wired, and addressed, it's time to start programming!

Chapter 7 - Programming on a Hardware Controller

The Tactile Approach

Different hardware DMX controllers vary wildly in the buttons available and naming conventions used. I have attempted to use the most generic, widely-recognized terms in the following chapter that will apply to the greatest number of controllers. Before sitting down to program, it might be useful to read this chapter with your controller's manual next to you. Reference the page in the manual that labels all of its functions and make sure you understand what each one does.

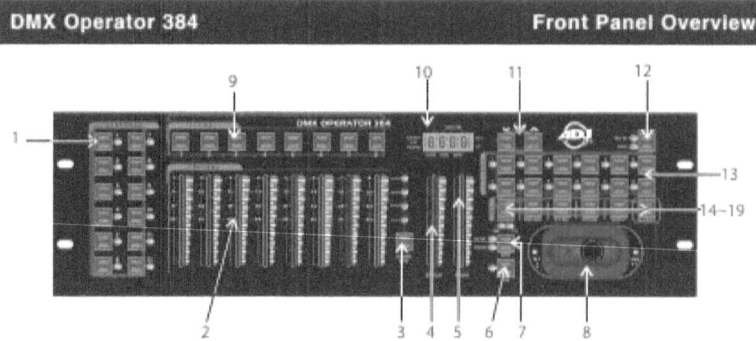

A page from the manual for an American DJ DMX Operator 384. A similar page in the manual of your controller can help you learn its individual controls and functions.

After you have set all of your lights to the correct addresses and turned on your controller, you are ready to program. When a hardware controller is turned on, it normally enters a mode called "blackout" first. Blackout mode is exactly what you think it is; it cancels all signals going to your lights and turns them all off. I imagine that the default here is blackout because otherwise, you might get a nice face full of blindness when you first powered your rig up. All output to the fixtures is canceled until the blackout button is deselected. This button normally hangs out off to the right side of a DMX controller, but yours may vary. Again, Keep your controller's manual handy while we go through this chapter, as you can reference if something I say doesn't line up exactly to your specific console.

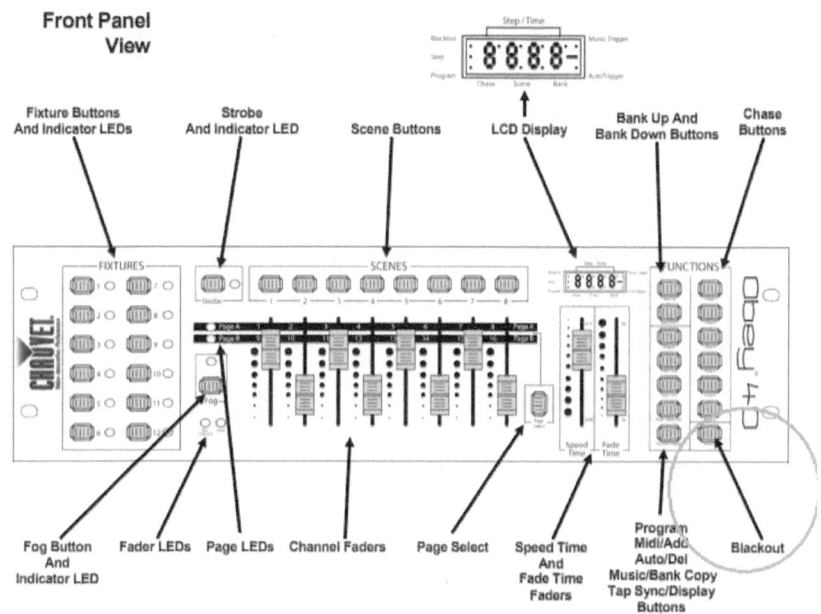

A diagram of one of Chauvet's simplest DMX controllers, the Obey 40. On the right side, I have circled the "functions" section, which contains the blackout button.

Manual (Playback) Mode

When you first disengage blackout, you enter manual control mode or playback mode. In this mode, you can select fixtures on the left side of the controller and adjust the sliders to see the lights respond in real time. Usually on the left side of the controller is the bank of 12 or 16 buttons that represent your fixtures. How do you know which button controls which light? Just think about your addressing. Button 1 will control the fixture(s) addressed to channel one. If your controller has 16 channels per fixture, button 2 will control the fixture(s) addressed to channel 17. To get a feel for programming, select a fixture and manipulate a few of the faders.

Find whichever fader is the master dimmer (usually dimmer one, but check that manual) and bring it up to the top (100%). Then find a fader that controls a color and slowly bring it up. You can play with the different color faders to create some unique colors through mixing. If your fixture has a channel for strobe control, move that fader up and observe the effects. If you adjust your faders and then deselect the light with its button on the left, the fixture will stay where you left it. This means you can move the sliders now and nothing will happen. If you reselect the light and move the faders again, the light will respond. If you select a different light and move the sliders, you can layer that light on top of the first without affecting it. Always remember to deselect a light once you have it set to your liking; otherwise the moment you move your sliders to adjust the next light it will mess up the first!

Scenes

So, all this adjusting is well and good, but unfortunately, in manual mode, none of your beautiful creations will be saved when you turn off the controller or change your sliders. For that to happen, we need to enter **programming mode**. Find the button on your controller marked **Program** and hold it down for a few seconds (you should get some type of visual feedback on the LED-display like a flash or an indicator light). Now you can select the fixture(s) you want to adjust and "set the scene" to your liking with your sliders. Don't forget to deselect each fixture before moving on to the next one!

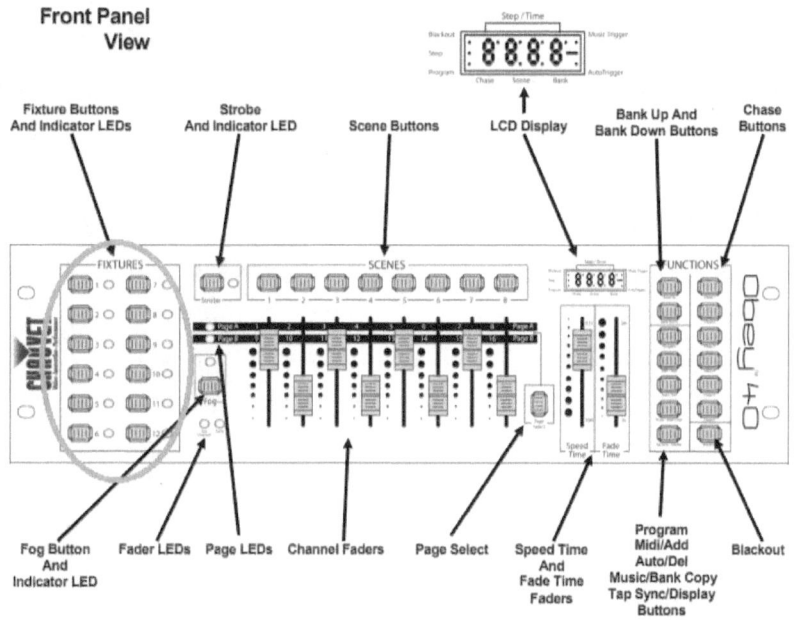

The Chauvet Obey 40 again, this time with the fixture selection buttons on the left side circled. Each button has an LED next to it which tells you which fixture you are currently programming.

Once you are satisfied with the look of your scene, you'll want to press the **Midi/REC** button (or whichever button is used to save your scene). Immediately after pressing record push one of the numbered **Scene** buttons at the top (or wherever) of your controller. Once again, you should receive some type of visual feedback like a flashing light. You've just saved your first scene!

Once you have filled up all of the scene buttons on your controller with your beautiful creations, you can use the **BANK UP** and **BANK DOWN** (sometimes called **PAGE UP/DOWN**) to move to another page, giving you a fresh bank of buttons to program onto. Don't make the mistake of recording one scene over another; this will delete your progress! Whenever you need to record a new scene make sure you do it to a new scene button. You can repeat this process of setting scenes and saving them to scene buttons until you have as many different scenes as you would like. Once you have finished creating some scenes (you can always come back and create more), hold the **Program** button again to exit program mode. Your controller should automatically jump back to blackout mode. If you go back at another time to program more scenes, just make sure you have a note of where you left off in your scene bank so that you don't overwrite your other scenes. If you read a little more in your controller's manual you can become familiar with other functions that have to do with scenes such as editing scenes, copying scenes, and deleting scenes.

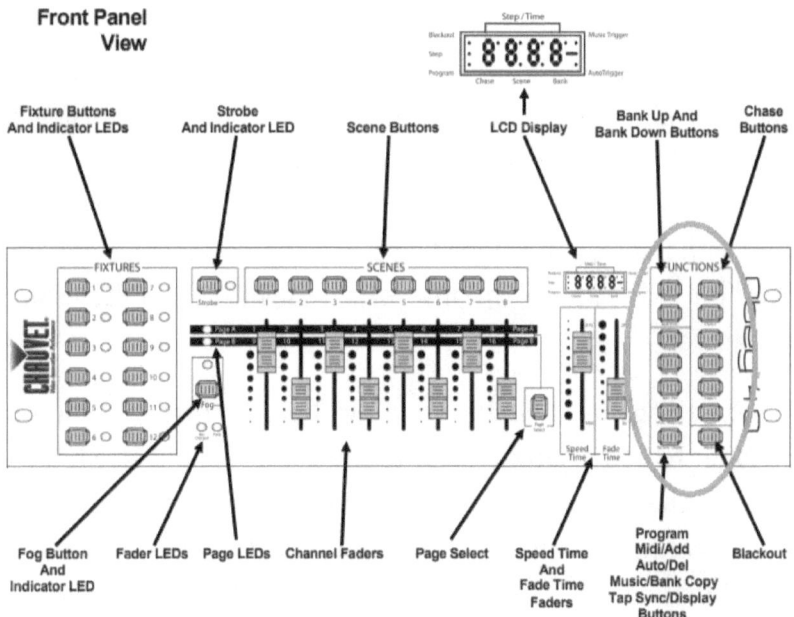

The "functions" section of this controller is super important because, in addition to the blackout button, it contains the buttons for entering programming mode, saving scenes, deleting scenes, and all of the buttons used to store chases.

Example and Methods for Scenes

I know that some people do better with actual walk-throughs, so let's get our imaginations going. First, I would have my manuals for each light close by, so I can see what each fader will control. I usually start with my wash lighting, as this forms the base for my light show. I select fixture one (or whatever I programmed them to). Remember that address one could be a single wash light or a group of multiple identical fixtures. I would bring up the main dimmer slider to full on, and then bring up the red slider so that my wash lights were on full red. Next, I would deselect fixture 1, so that any movements with the faders I make next won't mess with the red I just set.

Now I would select fixture 2, my (imaginary) scanners. I would bring their dimmer up, and then move the fader controlling the color wheel until they are red as well. I might move the pan and tilt faders to aim my scanners on the dance floor or ceiling. Am I happy with the way they look? If so, I would deselect fixture 2 and move on to fixture 3, my effect lights (let's say some moonflowers).

To shake things up, I would bring up their dimmer and then the blue fader to add a second color to my scene. I could also bring up the fader that controls rotation and add a slow rotation to get some movement in my scene. Most moonflowers also have a channel that controls speed, so I could use that fader to make the rotation faster or slower. After deselecting fixture 3, I can hit the midi/record button and then the button for scene 1 to save the scene. Congrats! We have our first scene, a red and blue scene with a little bit of movement.

Summary: A scene is created by selecting a light or groups of lights and adjusting the attributes (brightness, color, movement) to your liking. The first light is deselected and then the next light or group of lights is programmed in the same way. Once all lights are programmed, the scene is saved (recorded) to a scene button.

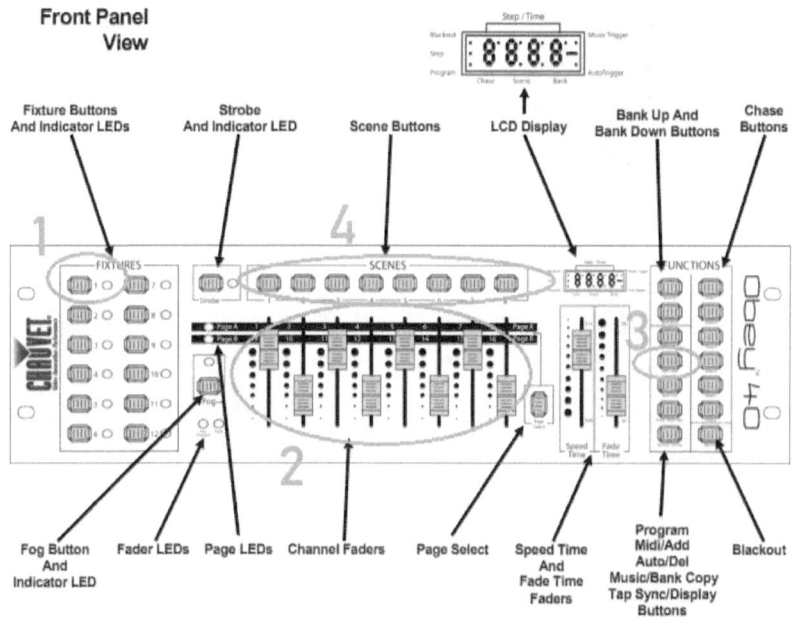

A visual representation showing the typical programming workflow: 1) Selecting a fixture 2) Adjusting channels with the sliders 3) Pressing midi/record button 4) Pressing numbered scene button to save the scene

I do have a workflow that I use when programming a general show for use at a school dance or wedding, which I am going to share with you! It goes a little something like this. I'll usually program a few simple scenes like the one above with slow movement and static colors that I can use for song breakdowns and slow dances. There is nothing more annoying than slow dancing with "rave lighting" going on, so by having some slow scenes I can create an atmosphere conducive to the song (which is the whole point of lighting control). I also program some scenes with medium-fast movements, which are great for hip-hop songs and other songs around 80-110 BPM. I try and introduce variety by having some scenes with a single color and some with dual colors (tricolor scenes happen very rarely). Finally, I record faster scenes to use for dance music, EDM, faster pop, and anytime I want to amp the energy up. Normally, I use cooler colors (blue, purple, pink) for slow scenes and warmer colors (yellow, red, white) for fast scenes.

*Color theory, or choosing colors that evoke certain emotions or are tied to certain situations, is a very interesting topic. I cover it a little more in my first book, "Building a Mobile DJ Light Show: The Essential Guide." There is also a lot of information on this subject on the internet, so go enhance your knowledge!

Chases

You may want to combine multiple scenes together to create a sequence that you can use for the duration of a song or group of songs. Once again, hold down the **Program** button to enter programming mode. Select one of the **Chase** buttons and it should light up or give some indication it is selected. Next, think of the first scene you would like to include in your chase, navigate to it, and select it. Hit that **Midi/REC** button and it will record the scene as a step. Navigate to or select the next scene you want to use, hit **Midi/REC** again, and repeat to add more steps. Your manual will let you know exactly how many steps you can add to each chase. Hold **Program** to leave programming mode and save the chase. Some controllers even let you do a bulk add a whole group of scenes to a chase.

Example: The most basic and common chase you can create is a color chase. Let's imagine I created a red static (non-moving) scene and a blue static scene. If I'd like for those two colors to alternated back and forth with each other without having to constantly press their scene buttons, I could create a chase. Using the procedure above, I would save those two colors (and any more, if I'd like) to a chase button. Once I'm out of programming mode, selecting that chase button will cause those colors to alternate automatically. You can adjust the "speed" and "fade time" sliders to control how quickly the colors switch and whether that fade together or snap instantly from one to another.

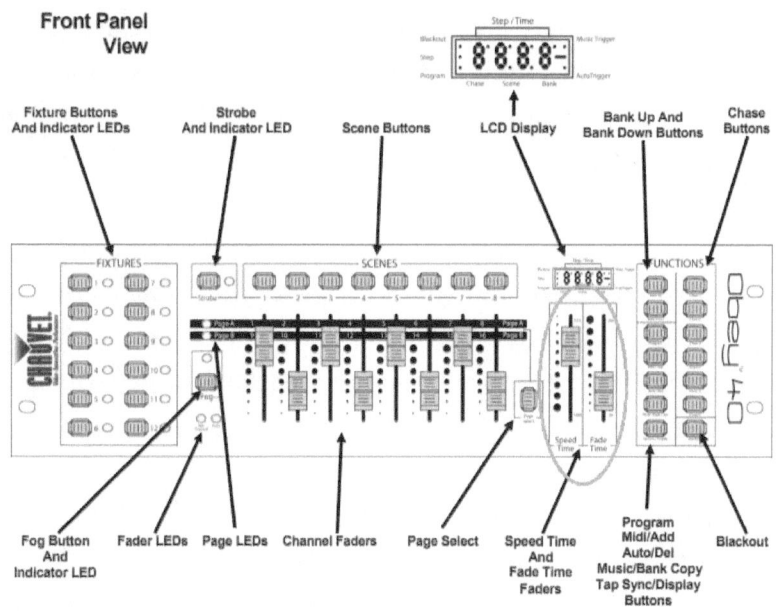

In the diagram above, the speed and fade time sliders are circled. These affect the chases you create.

Macros

One of the best ways to program your lights quickly is by using the light's built-in **macros.** Macros are color chases, dimmer fades, and movements that were pre-programmed by the lighting manufacturer at the factory. Usually, they are kept on one of the channels in the light and can be accessed by bringing up the dimmer and then the "macro" channel fader. Now, instead of you programming multiple scenes to create color chases or complex movements, you can simply use the built-in macro effects and take advantage of someone else's programming.

Experiment with your lights and see what macros are available. Select a fixture, bring up its dimmer, and very slowly slide the channel fader up that corresponds to the light's macros to cycle through the different programs. Macros are great resources, but keep in mind that they are the same programs you can access in the automatic mode. They work in a pinch if you need some quick scene changes, but you will most likely want to be creative and create your own scenes manually with the other sliders.

Playback

Now that you have created some scenes and possibly chases, you can play them back live at your event. Remember manual mode, the mode you enter when you deselect blackout or leave program mode? Manual mode is where you will trigger your scenes and chases. Simply turn off blackout and select the scene you want to use. BOOM! There are all your settings, recalled and running. If you want to switch scenes just hit the next scene or chase button. Scroll through the scene banks just like in programming mode to find your other scenes and trigger them.

If you want your scenes to progress automatically, you can select a button most controllers have marked **Auto**. Now the controller will progress through the scenes in the scene bank automatically, and you can change the rate of these changes with the sliders marked **Speed** and **Fade**. If your controller has a button marked **Tap**, you can tap that button to the BPM of your choosing to set the rate that the scenes change at. Some controllers even have built-in microphones and selecting the **Music** button will cause the controller to progress through the scenes to the beat of your music. Because the quality of this microphone can vary, as can the volume and bass of your music, this mode may or may not work well for you.

The playback options I discussed in the previous paragraph can be highly variable from controller to controller and would again urge you to familiarize yourself with your unit's manual. Most will offer, at a minimum, the playback controls I listed above, and many will offer even more!

A Note on Moving Heads & Scanners

There is one topic I need to discuss particularly as it concerns hardware DJ controllers, and that is the programming of movements for scanners and moving heads. In some of the most recently released lights from manufacturers are specific channels with movement patterns pre-built into a specific channel (like the macros we discussed earlier, but without colors and other effects). This is a great thing, as programming movements into these intelligent fixtures on a hardware controller can be difficult.

The process for creating movement with moving heads and scanners on a basic lighting console goes something like this; create multiple scenes on your controller with your moving heads or scanners aiming in a different direction in each scene. Then, combine those scenes in a specific order into a chase button. Then, during playback, adjust the fade and hold times (look for these buttons/sliders toward the right side) until you get as smooth of a movement as possible. Sounds simple, but in actuality can take a LONG time to perfect. This aspect of programming on a hardware controller was one of the biggest factors in my decision to switch to a software lighting program.

Conclusion

There you have it! Those are all the basic components of programming lighting on a hardware DMX controller. It may take some time to get used to the functions and processes, but with time you can develop some slick muscle memory to speed up the process and become proficient at whizzing along while programming. If you program on a software program and read this chapter anyway, read on to see the differences in programming with software.

Chapter 8 – Programming on a Software Lighting Program

The easiest way to do it

If you jumped straight to this section because you don't own a hardware lighting controller, that's fine! However, if you are new to the concept of lighting programming and have never attempted this before, I suggest you go read chapter 7. Many of the concepts are similar, and doing them in software is simply easier. If you already read chapter 7, you're ahead of the game!

The whole basis of lighting programming within a lighting program is the same as with a hardware controller. You are going to adjust the DMX channels of your lighting, create scenes and chases, and then play them back during your event. This process is **create, record, and replay**. Before we ever get to the point of being able to program our light show, you'll need to correctly install and set up your lighting software. Regardless of the company you choose, the process is usually similar to installing any other software (check your manual for instructions).

Once you have the program installed and your computer restarted, plug in your DMX dongle. Give it a few seconds for the computer to recognize the hardware; I often will plug in my dongle and then go work on some other aspect of setting up. After a few seconds, you can start the program. You're most likely going to be greeted by a screen similar to the screenshot of MyDMX below:

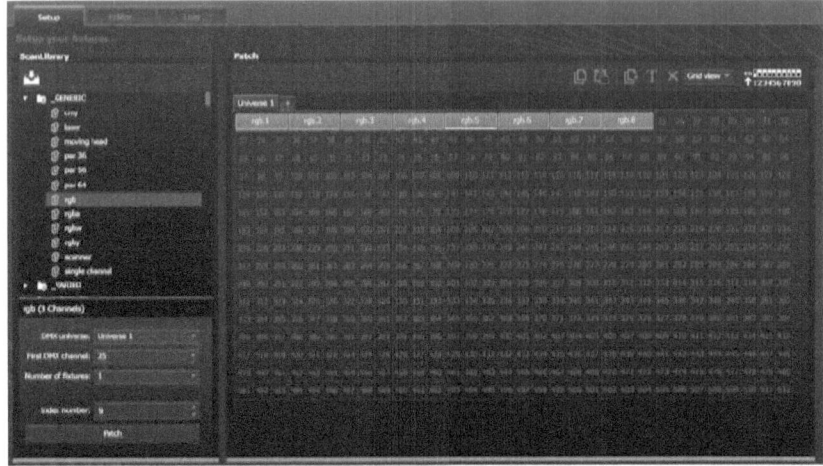

A screenshot of MyDMX. The setup, editor, and live tabs are found in the upper left.

Take a look at your own software; the basis of the software revolves around the three tabs in the upper left of the screen. Usually, there is a setup or fixtures tab, an editor tab, and a live tab. We are going to walk through each tab and its purpose.

A screenshot of the main screen of ShowXpress. Within the circled area are buttons that allow you to access the various programming and playback screens.

Setup

In MyDMX, the setup tab is clearly labeled. In ShowXpress, you'll need to click the editor tab at the top and then the "Fixtures" tab beneath it. The setup tab is where you will import the profiles for your lighting fixtures.

If you remember back to earlier chapters, a lighting profile is a pre-made file that knows what each channel of your light does and displays it on the screen. This saves you from having to look in a manual during the entire programming process. In MyDMX, there is a taskbar on the left with a list of lighting manufacturers. Simply find your lighting manufacturer and then the specific model of light you have in that list and then drag it into the large section on the right filled with numbers. Those numbers represent all of the channels available for you to fill with fixtures. Wherever you drop the profile will determine which channels your light will take up. You can then go to the back of your lights and address them according to the channels you just assigned them (remember the DMX starting address is the first channel your light uses).

You can also "patch" (the word used when assigning lights to certain addresses in software) in your fixtures with the dropdown boxes below the fixture list. It allows you to add multiples of the same fixture into the software at once and select the start address for the fixtures.

In ShowXpress there is an "Add Fixture" button right at the top left of the large numbered grid (hover over the icons to find it). Clicking this will bring up a similar list of lighting manufacturers which you can navigate to your specific light model. Once you have selected the light you want, select the number of fixtures you would like to use and the starting address for the first one and then hit patch. You will see a bar (or in the case of more than one light, multiple bars) stretch across the number grid, representing your light and the channels it is occupying. Repeat this process to add all of your fixtures to the large numbered DMX grid.

One great feature in ShowXpress is the 2D viewer to the right of your DMX grid. As you add fixtures, it will be populated with small icons and pictures representing your lights. You can drag and arrange these little pictures to create a graphical representation of your light layout on your stand or truss. Not necessary, but make selecting lights easier down the road.

Summary: Add your lights to your programming software first, then physically address your lights to the selected channels by referencing your software.

Editor

Once all of your fixtures are imported into the program, it's time to edit them and create your scenes and steps. Once you switch over to the editor tab, you will see a large bank of sliders along the bottom of the screen that represent the physical sliders found on a hardware DMX controller. Remember, each slider controls one DMX channel or one parameter of your lighting.

What makes lighting software programs so much more convenient to use than hardware controllers are the small descriptions and icons above each slider that tell you exactly what each channel controls. At the top of the slider is the channel number (starting at 1 and going through 512). Below the channel number is a number representing the value that the slider is currently at (0 through 255). Additionally, you will have a small bit of text that serves as your description of the channel (hovering over this text usually brings up any text that may be cut off). Clicking in the space below the text will bring up a list of ALL the functions of that channel.

Example: A certain channel controls the color wheel of a moving head light. Instead of clicking with the mouse and dragging the slider (which you can still do), you can simply click in that space and select the color you want. 2 clicks and you have your channel set! This can be done much easier than consulting a manual, finding the value that you want, and then sliding a physical slider to just the right position.

*A quick note: ShowXpress has an option to turn off the DMX signal your dongle is sending to your lights. If you are adjusting the channel sliders within ShowXpress and your lights are not responding, you will need to make sure you are sending DMX signal from the editor screen. In the top right-hand section of the

screen is a small icon that looks like a DMX/XLR connector with either a red or green bar beneath it. When the bar is red, you are not sending DMX signal out to your lights. When the box is green, the signal is being sent. If you are making changes and do not see your lights responding, always double check this icon! It will appear on most screens in ShowXpress. It's there to prevent you from sending multiple signals simultaneously from different areas of the program. Only activate it in whichever section you are currently working in, and make sure to deactivate it if you switch screens (say, from editor to live).

Depending on your software, the other sections of the editor screen will look different. One thing that all software programs will have in common, however, is the box that contains **steps**. These are the scenes that you create and save, just like with a hardware controller. If the box contains a single step, it is static and similar to a **scene**. If you add multiple steps to this box, it becomes similar to a **chase**. Don't worry about the other sections or details for now; the steps are what we are going to focus on first.

Steps (Scenes)

In similar fashion to the creation of scenes on a hardware controller, you can create scenes within a lighting program. Simply click the virtual sliders at the bottom and drag them to the values you want. You can also use the method described earlier of clicking in the space between the slider and the description or icon and selecting the value or attribute you want for that channel. Try running through the example given in chapter 7 (wash lights, scanners, and moonflower) to create the same effect. Once you have your scene set how you would like it, click the **save** icon or simply push ctrl+s (just like saving a word document). You can type a filename for the scene and save it for later. I recommend using descriptions of the scene for your filenames, such as "slow red" or "quick blue chase." Congrats! You've just created your first one step scene.

Bonus Tip: One benefit of lighting software is that even if you don't own the fixtures you can still play around with them in the software. If you are renting lights for a large production or anticipating a shipment of new fixtures you can add them to your software and being programming them before they are actually connected to your dongle.

Another Bonus Tip: It's up to you how you choose to name your lighting scenes. I recommend making the title descriptive of the scene (as mentioned above). Additionally, use abbreviations wherever possible. A scene filled with green wash lighting could be named "WSH.GRN" and a scene made specifically for a slow dance could be called "SLW.DNC"

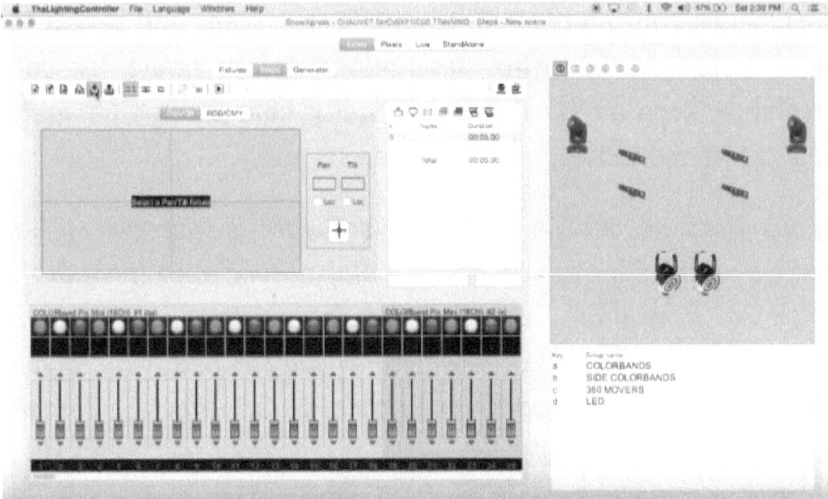

A screenshot of the ShowXpress scene programming page. You'll find your programming sliders in the bottom left of the screen, a visual representation of the lights in the top right, and the "steps" box just to the left of that.

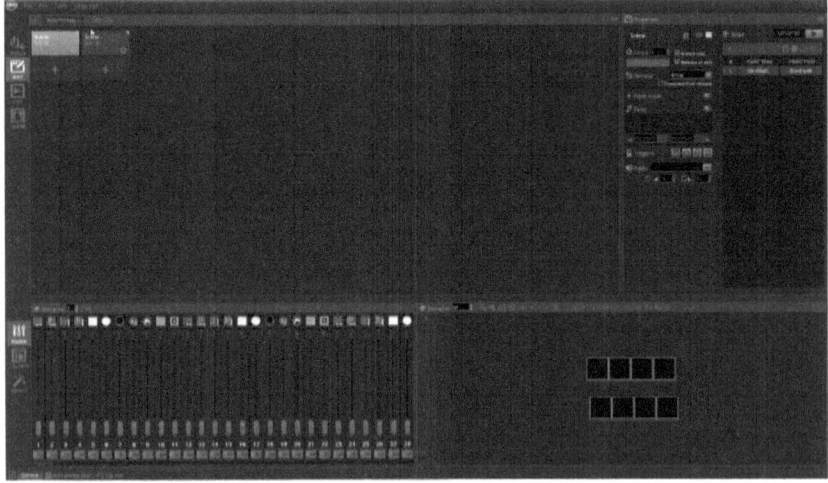

A screenshot of MyDMX 3.0's programming page (called "edit"). Notice your bank of sliders in the bottom left of the screen and the collection of saved scenes in the large box that fills the top left of the screen.

Now let's try a multistep scene (similar to a chase in a hardware controller). All you do is set your channels how you would like (let's say to red, full brightness) and instead of hitting save, go over to your "steps" box. You'll know you're in the step box because you will see a number list with times (00:00:00) next to it. Click the "add step" icon (usually involves a plus symbol, so hover over your icons until you see "add"). You'll see another step pop up below #1. Now you can go make some changes to your lighting channel sliders (how about setting them to blue?).

If you go back to the steps box and click step #1, your lighting will go back to red. Clicking step #2 will go to blue. With step #2 highlighted you can click the "add step" icon again to add another step. Make some changes (let's try green) and then click through your steps again to see them change. You can adjust the "hold" or "duration" section of the step to change how long each step will last in the sequence. Look around the different taskbars for a button that looks like a play button (right-facing triangle, sometimes in a box). If you click that button, your steps will begin to play in order. Click the play button again to stop playback and continue editing if you would like. Continue adding steps and adjusting their duration. Once you are satisfied with your multi-step scene, hit save and name your scene. That's all there is to it!

Right here we are going to separate into two small sections to discuss a few differences between ShowXpress and MyDMX.

MyDMX

You have some additional options available to you within the Steps box in MyDMX. Adjusting the value of the "Fade Time" will change the amount of time that it takes your steps to blend into one another. There is also a dimmer bar that can adjust each step's master brightness (a cool feature in my opinion). MyDMX has one additional box besides the steps box, the **Scenes** box. Instead of saving scenes in a folder like ShowXpress, MyDMX puts all of your scenes into this box to access quickly. Whenever you are finished with creating a scene, simply click the "add scene" icon (which looks like a paper with a plus sign) to start another. You can rename the scenes with useful phrases such as "Static Red," "Fast Blue," "Rainbow," or some other convention that makes sense to you. To the right of the name of the scene, you have the option to choose how many times each scene will loop. The default setting will cause each scene to continually loop when activated, but this can be changed to a single loop or multiple. This is helpful if you have, for example, a scene that flashes or strobes that you don't want to remain on.

ShowXpress

ShowXpress has some additional fun features to make programming a breeze. You'll notice the 2D viewer has stuck with us from the setup screen and can be found off to the right. Here, you can select fixtures (use ctrl to select multiple) and edit those fixtures by themselves. If you select moving heads, a box to the far left will show a grid with a crosshair through it. Here, you can click and drag to adjust the X/Y position of your moving heads or scanners as opposed to using the pan and tilt sliders.

If you select a fixture with RGB LEDs, the box on the far left will show a color wheel that you can use to select colors easily and quickly. Another useful feature is the creation of groups. When you have selected multiple fixtures in the 2D viewer, right click and select create a group. You can name the group (for example, wash lighting) and assign it a hotkey. Whenever you press that key, those lights will be automatically selected. This makes for quick transitions between fixtures as you program.

Below each slider in ShowXpress is a little drawing of a line. It can either be a straight line with 2 plateaus or a curved line. Click this line to adjust the fade state of the channel. A smooth curved line will cause the steps in your scene to fade into each other, and the straight line will cause steps to snap one after another.

Live

The live tab is where you will play back the scenes and chases you created in the editor tab. Each software treats this section differently, so once again I'm going to separate for a moment to talk about how each functions.

MyDMX

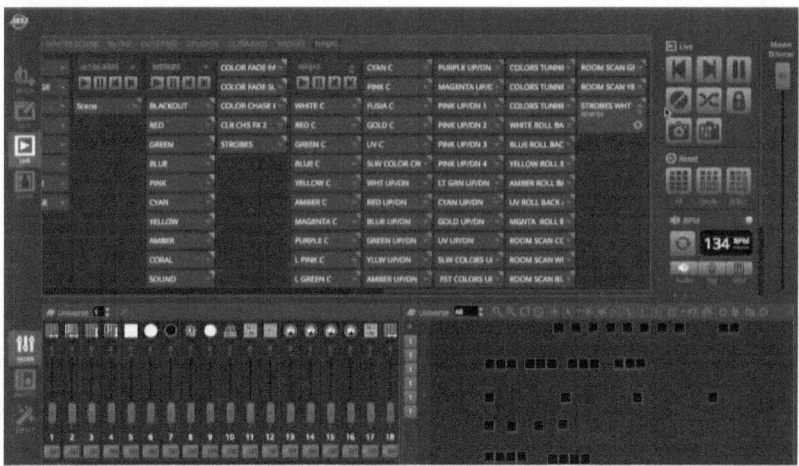

The live playback section of MyDMX. Look at all those scenes!

Scenes are automatically added to the "Live" section of MyDMX whenever you create them in the *Scene* box in the editor. If for some reason you create a scene that you *don't* want to see in your live view, simply click the checkbox to the left of the scene name. **Only the checked scenes will appear in the live tab.**

Once you enter the live tab, there are a few different options available to you to trigger your scenes. The first and easiest method is with your computer mouse or trackpad. Simply click on the scene you wish to playback and it will activate. If you chose to have the scene loop only once or any number of times, it will pause when it comes to the end of the loop.

Another option for live control is keyboard triggering. This is the preferred method of live triggering because it allows you to have a fast, responsive way to switch between your scenes. At your event, you will want to be able to call up scene changes quickly, either with changes in your music or activities. To program a keyboard trigger, simply right click on the scene and select "learn keyboard command," and then push a key on your keyboard. Now, whenever that key is pressed, your scene will activate.

When programming keyboard triggers it is always important to think of your organization. For me, I've found the best organization to be in rows and columns. I program my slowest or static scenes along one of the top rows of the keyboard, increasing the speed of the programs as I move down. Using this method, I know that no matter which key I trigger in a row I will have a predictable speed. Another method is the color method. When using the color method, I program an entire group of scenes of varying speeds in the same color and assign them to a group of keys. Now I can trigger a certain color and vary the speed with the changes in music or mood. Be creative in the layout of your keyboard triggers!

If you've purchased the full MyDMX software and not the buddy, you have the option to use MIDI control. We are going to discuss midi control in greater depth at a later point, but for now, I will explain how you can program MIDI commands. With your MIDI controller plugged in, right click on a scene in the same manner as when you programmed a keyboard trigger. Click "learn MIDI command" and then press a button on your MIDI controller. Now your scene is connected to that MIDI button and will be triggered just like a keyboard button. Sliders are also MIDI controllable if your controller has knobs or sliders. Simply right-click the fader, click "learn MIDI command," and run the knob or slide through its full range of motion.

One last important aspect of the **Live** tab is the ability to make modifications to your scenes on the fly. This can be especially helpful if, for example, you need to raise the height of the movements of your moving heads above the crowd or dim your scenes to not blind guests in a small room. Below each channel fader in the **Live** tab is a small box that by default says "Auto." This means that the fader is simply responding to the commands you programmed for the scene back in the editor tab. Clicking the "Auto" button will change the mode of the slider to "LTP," then "HTP," and then back to "Auto." **LTP** stands for latest takes priority (or precedence). This means the slider will respond to the last command it received, meaning if you click and move the slider with your mouse whatever you set the slider to will take priority over the scene programming. **HTP** stands for "highest takes priority" and means that the command with the highest value for that slider receives will take priority. If the scene already has a certain slider at the max value, you'll notice that you aren't able to move this slider down. However, if you have a scene where this fader is lower (such as a low-level movement of your lights across the crowd), you can raise the slider and it will stay in the raised state. Personally, I find **LTP** to be the most useful setting; it allows me to have control over the brightness of my lights live, giving me the option to program my lights at a single brightness setting and adjust it live for the event's needs.

ShowXpress

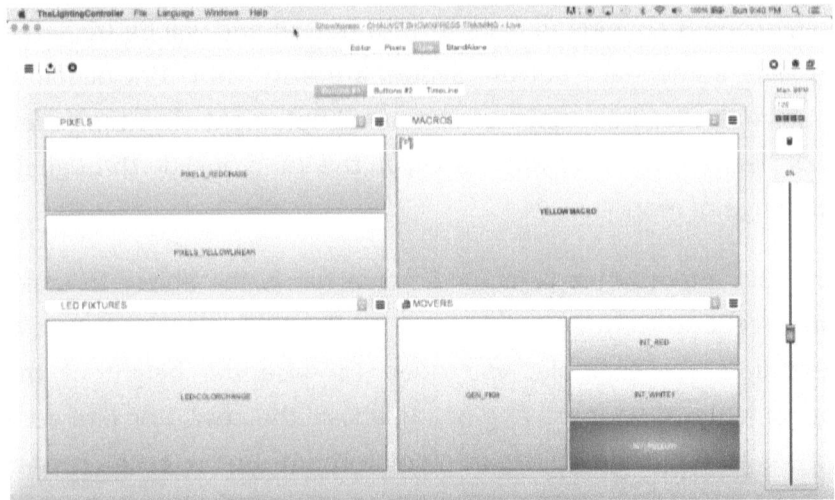

The live playback section of ShowXpress. The layout of scene buttons is very customizable!

Instead of having all your programmed scenes clumped into one giant section, the "live" tab of ShowXpress allows you to create multiple **Pages** within the window and fill them selectively with whatever scenes you choose. By default, you will have a single page with no buttons. Let's create 4 pages to start with. To the right of the drop-down box that says "Page_1" is a small button with a menu on it. Clicking it will bring up a small list of options, and you can select "Add Page" to insert an additional page. You can repeat this until you have the number of pages you need. I suggest you start by building a page for each type of fixture that you have. You could build a "wash" page, and "movers" page, and an "effect lights" page to name a few. You can use the same menu to rename or delete the pages if you choose. Clicking the drop-down menu on the bar that contains the page name will let you switch between the pages you have created. That's great, but you're going to want to be able to have multiple pages on your screen at once.

To do this, click the small menu button to the top left of the page section for "settings." Once you enter the settings box, you can change the number of "boards" in the first tab section to 4. Now you have 4 different drop-down menus where you can select the 4 pages you made earlier. This process can continue for as many boards and pages as you would like to make. You can even create multiple tabs that each contain multiple boards. You can also open a free-floating "child window" that will give you even more boards and pages (this is a more advanced feature – consult the ShowXpress manual for more information).

To add a scene to one of the boards on your screen, just hit the same small menu button to the right of the page name and select "Add Light Scene." This will open a file explorer where you can select the scenes you made earlier that you would like to import. This brings up an important point; **keep track of where you save your scenes when you make them in the editor!** Use ctrl+click to highlight multiple scenes. Once you import them, they will appear within that board. As long as you have the DMX signal icon in the top right corner activated (green bar), clicking on a scene button will turn the scene on. Simply repeat this process within each board to add different scenes to each board.

Now seems like a great time to bring up one of the most important and cool aspects of the ShowXpress software, and that is the ability to **layer** your scenes. Click a few of the scene buttons you have added to your boards and you will notice that you can select multiple scenes at once. If you are watching your lights you might notice some crazy behavior. This is due to ShowXpress allowing you to layer your scenes one on top of the other. Why would this be important? Because it allows you some amazing flexibility when you are programming your show live. Here's how: if you have moving heads, program some scenes that **only** contain movement, some scenes that **only** contain the dimmer, and some scenes that **only** contain colors. **Don't touch the other sliders when you are programming these scenes so that they aren't activated.** Save those movement, color, and dimmer scenes and then import them into the "live" tab. Now you can click to activate a movement, a color, and a dimmer scene together and you've created a unique live scene! You can change the color and the movement and brightness will continue without interruption. You can also switch the movement and the color and dimmer will not be affected. Now, instead of having to program a million different color/movement/dimmer combinations, you can make a few base movements and colors and combine them all live. It saves a lot of time and gives you a lot of on-the-fly creativity!

Example: Imagine you're programming on a hardware controller. You've made a beautiful blue scene with your moving heads gliding slowly around the room. What if you love everything about the brightness and speed of the lights, but you want all of the lights to be red instead? On a hardware controller, you'll have to change the color sliders for all of the lights and save the scene in a completely new slot. You have the two scenes you wanted, but they're occupying two of the precious spaces in your scene bank. And what if you want the same scene in green, yellow, white, and cyan? You'll need to create and save a scene for all of them. Want to change the speed? You'll need a new scene for that too. Bummer!

In ShowXpress, you program the movements, colors, and

brightness all as separate scenes. While this may initially take more time, you will have increased flexibility when it comes to your event. You can recreate the blue scene with slow-moving lights and then, with the press of a button, change the color to red, green, yellow, or any other color you choose. You could also leave the lights blue and select a button for a different movement or brightness. By layering different components of a scene, you have unlimited flexibility without creating hundreds of different scenes.

Other Live Options in ShowXpress

There are many other parameters and options you can control for each page or board in the live section. Here are two that I use quite frequently that you will most likely find useful.

Clicking the small menu button next to a page, look for the "solo buttons" option. Selecting this option will cause only one of the buttons within a board to remain active at once. If you have already selected a scene and then try and click another scene within the same board, it will cause the first scene to turn off as the second scene turns on. This is useful if you have a board dedicated to movements, as it will prevent you from layering multiple movements and causing your lights to go crazy.

If you right click on a specific scene within a board, you'll see the "flash button" option. By selecting this option, your scene will become (you guessed it) a flash button. This means that, when you click to trigger the scene, it will only stay activated for as long as you hold down the mouse or keyboard trigger. As soon as you let go, the scene will turn back off. This is especially useful for a strobe scene, as you wouldn't want to leave your strobes on for an extended period of time. It also allows you to easily time strobes or flashes with the beat.

Chapter 9 – Scene Builder and the Generator

Making your life easier... SO much easier

Following the programming steps in the previous chapter for programming in a lighting program is all well and good. It's simple, straightforward, and easy to follow. But let's be honest; what's the point of dragging and clicking virtual sliders to program a light show? That's the exact same thing we did on a hardware controller! Didn't I tell you that software would be quicker and easier? Don't you worry my friends, because it is. It's time to introduce you to the best features that lighting software has to offer.

These features, called the "Scene Builder" in MyDMX and the "Generator" in ShowXpress, speed up your programming 100-fold. They allow you to create complex effects with just a few clicks of your mouse, saving you the tedious work of creating and adjusting hundreds of steps within a scene. The process of using them is similar, but not identical. If you are using a program other than MyDMX or ShowXpress, you may or may not have a feature similar to the generator. If you do, and you can figure it out, more power to you. If you don't have a generator or scene builder in the program you are using, I **highly** suggest that you switch programs. Once you understand how to use the generator and scene builder, you will rarely use the virtual sliders to program again.

If you have any experience with web design, you'll understand that the scene builder and generator work like a visual website builder. Building a website normally involves long lines and paragraphs of code that, while easy to understand and write to a programmer, can be mind-boggling to the average internet user. A visual website builder allows you to drag and drop different website elements on a website on the front end of a website, while a computer program writes and adjusts the website's code in the background.

With the generator and scene builder, you are able to create effects such as moving head patterns and color fades by dragging or clicking your mouse (depending on your program) while the program creates and times the steps of the scene automatically in the background. It's really neat stuff and makes ALL the difference in programming time. Let's take a brief look at the two systems.

Scene Builder – MyDMX

Head to your editor tab and create a new scene (icon with a page and a + sign above the scene box). Then click the **Scene Builder** button in the very top taskbar, directly to the left of the circle that says 3D. A new window will open with 2 sections. On the left, you have a visual representation of your lighting fixtures in 2D space. Each of the small squares you see in this area represents either a single light or a pixel section of a light (if your light has multiple LED sections that can be controlled). If your light has multiple sections that can be controlled, such as a LED bar, it will have multiple squares hooked together. You can hover over a square to see what it is and move it around by clicking and dragging. Now would be a good time to arrange these squares identically to your actual setup and create a visual representation. There are other tools in the taskbar above to help you create this visual representation, such as buttons that will evenly space your lights across the screen. You can come up with your own organization for your lights; I usually opt to group similar fixtures together, such as those that have RGB LEDs and those that have color wheels.

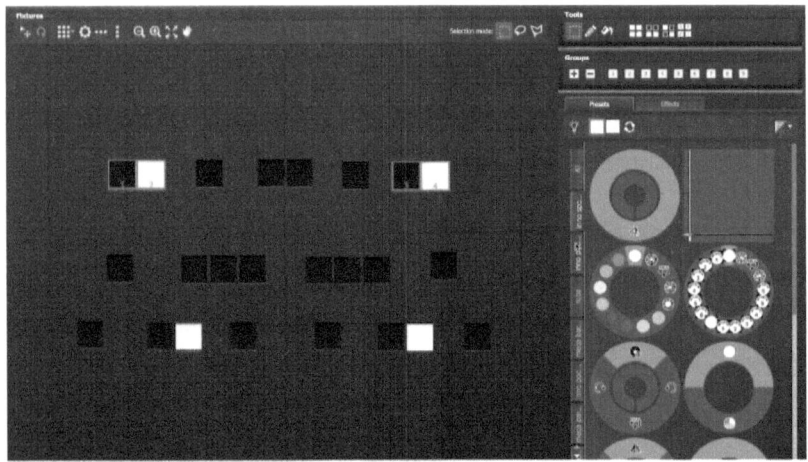

The scene builder within MyDMX. The squares on the left represent all of the lights you are programming. The circles on the right allow you to quickly pick colors and gobos, adjust brightness, and much more (without moving any sliders manually).

In the right section are a bunch of controls that allow you actually create the various effects. You'll notice a bunch of tabs on the left that are broken down by fixture type. Selecting different tabs will bring up the specific options available for that fixture, such as a color wheel or a movement grid. These wheels are great; you can quickly and easily select a gobo, color, brightness, and strobe in a matter of seconds. If you want to see these changes, you need to turn on your lights. To do this, select the fixtures you want to turn on with your mouse and click the small button that looks like a lightbulb (found directly above the color and options wheels).

Continue to explore the buttons above this right section, as they allow you to select different fixtures and create groups that will speed up programming. If you would like to have multiple steps in your scene, first create the look you would like for the first step in the scene builder. Then minimize or move the scene builder window to the side, click "add a step" in the steps box, and then open the scene builder again. You can now edit the second step of your scene and continue on in this manner.

So now you know how to create some simple scenes easily through the scene builder. But there is an even more exciting aspect of the scene builder, and that is the **effects engine**. This is the place where all your wildest dreams come true. Really, I'm serious. The effects engine makes me giddy. It's what is going to allow you to create complex movements and color fades/jumps in a matter of moments.

First, go to your scenes box, create a new scene, and then open the scene builder. In the right-hand section, click the tab labeled "effects." If you have moving lights, select those first. Click the small button with the "Fx" on it and a drop-down list will appear. Let's focus on the position and curve effects first. Click "position effect." The program will immediately create a circle on the screen and your lights will begin tracing that circle. You can adjust a TON of parameters here so I won't get into every single one, but do some exploring. You can drag and drop the points of the shape, change the shape type, the number of points, and the phasing value.

The **phasing** value will offset your lights' movements (if you have multiple moving lights selected) to create a cool "out of sync" look. Imagine we have two moving heads – with the phasing value set at 0, both lights will move in exactly the same way. As you increase the phasing value, one of the lights will become increasingly out of sync from the other. You can also use the "duration" box at the top to adjust the length of the effect, which will affect the speed. A longer duration means the light has more time to complete its pattern, and will move more slowly. A short duration will cause the light to move more quickly. Click the checkmark next to the duration box to save the scene, which will automatically appear in your scene box.

Another effect option you can choose is a curve effect. Creating a curve effect will cause a box with a curved line in it to appear. There are multiple curve options to choose from, and many parameters to adjust. However, the first thing you will want to adjust is the "channel(s) which are playing the effect." This will determine what attribute of your light the curve will control. For example, let's say you make a simple sine wave curve and in the drop-down box labeled "channel(s) which are playing the effect" you select "dimmer." Now the dimmer for the light(s) you selected will follow the curve you created. Cool huh? You've created an effect that would have taken a very long time to make otherwise.

The last couple effects options have to do with color and pixel mapping, and they are so much fun to use (I'm sure you will agree after you try them). Open up a new scene and then open the scene builder. Select some lights that are RGB color mixing lights such as wash pars or bars. Click the "Fx" button and select "pixel effect." You should immediately see the effect not only on your lights but on your screen. Likewise, you can select the "Fx" button and select "color matrix effect." There are a vast number of effects that you can modify and change; so many that there is no possible way I could explain them all in this book. Take some time to play around with all of the different options and effects. I've saved these effects for last because they are the most exciting and fun to play around with, so make sure you experiment with all of them!

The Generator – ShowXpress

The generator in ShowXpress functions similar to MyDMX's scene builder. You'll find the generator by first heading to the **editor** section and then selecting the **generator** tab. Once again, we will start with our moving lights. You'll see your 2D representation of your setup to the right, a box called the "curve box" in the center, and a box labeled "channels" on the left. Select a pair or group of moving lights. The curve box will fill with a grid and the channels box will populate with all of the channels shared by the selected fixtures.

We'll start with the pan and tilt channels, so click the checkbox for those channels in the channels box. This activates the channels. **You'll do this for any attribute you want to adjust**. The curve box should already contain a default large circle that fills the box. If you hit play in the taskbar (remember to have your green DMX icon active in the top right so that DMX signal is being sent) you will see your fixtures react and begin to trace the circle. You can click and drag the points of the circle to change the shape and make it larger, smaller, turn it into a figure eight, etc. You can also right-click on one of the points and add a point, delete a point, or reverse the order of the points, along with other options. This allows you to create all sorts of new designs and patterns.

Below the curve box are a few important options. Adjusting the duration will adjust the speed at which the fixtures will move through the shape. By default, the option labeled "curves" is selected below the duration. This will give you smooth curved lines between the points of your shape. You can change the option to "lines" or "points" and see the different effects it creates.

The last option you can change at the bottom of these settings is the "shift" percentage. Shift, like "phasing" in *MyDMX*, will create a delay between the movements of your moving heads. Instead of them all moving simultaneously along the path of the shape, they will move out of phase. The more you slide this bar left or right, the more out of phase your moving heads will be. Experiment with different amounts of shift to create effects that would have taken you hours to create in the "steps" editor. Don't you love saving time?

The generator is not limited to just movement programming. If you select a different attribute such as "dimmer" in the **Channels** box (don't forget to uncheck pan/tilt or whatever else was selected previously) it will populate your curve box with different options. In the case of the dimmer, it is a single line running across the top of the box with two points, one on each end. Right-click anywhere along the line and "add a point." Just like in the pan/tilt section, you can now click and drag your points to adjust the line. You have the same options available to you as before for modifying the curve and shape of the line. If, for example, you create a bell curve in the dimmer section and click play, you will see your light dim and re-brighten as the scene progresses through the duration.

Helpful tip: Most moving lights have a shutter channel, which basically functions as an on/off channel for the light beam. If you're trying out the dimmer curves and don't see your lights responding, you'll want to click the box for the shutter so that channel is activated.

You can apply these same concepts to any of the sections, such as color or focus. **A quick reminder:** If you plan on layering scenes such as colors and movements on top of each other later, make sure that you don't program any color, shutter, or dimmer values when you are programming movements (keep those boxes unchecked). If you do, all the dimmers and colors will combine and you'll have a mess! **Only modify the attribute you want, and then you can combine them later in the live screen.**

In Conclusion

While the information and walkthroughs I have provided here are more than enough to get you on your feet with lighting programming, there are many more features that these lighting programs contain that I did not touch on. These settings and customizations can take your lighting to a whole new level, but they are beyond the scope of this beginner's book. Each software has its own forum where users can discuss the features, problems, and tricks of the program, and I would encourage you to become a member and participate in these forums. A quick search in the forum will usually yield dozens of answers to any question you may have.

Chapter 10 – Putting it All Together; My Tips, Tricks, and Workflow

This is the chapter you've all been waiting for

Once you understand the basics of how DMX works, how to hook up your lights, and how to manipulate your lighting controller or software you feel like the king of the world. No doubt you probably have spent a few minutes (or hours) tinkering with different settings and playing around with fun effects and ideas. You've mastered the ins and outs of programming but you're at a loss for how to bring all of this technical knowledge together into an exciting light show that will evoke the emotion you or your client is seeking. After all, no one is hiring you because they want to see your "scenes" and your "triggering." They want the end product, the benefit, the emotion.

This chapter is all about my personal programming workflow. I'm going to discuss my thoughts on choosing colors, creating different movements and positions, judging the energy of a song and incorporating lights to it, laying out my keyboard or MIDI controller, and playing back the lights live. I hope that instead of copying my ideas exactly (which you are welcome to do if you so choose) you will use them as ideas that catalyze your own creativity. The joy of programming lighting is that YOU can create literally whatever comes to your mind as long as you have the equipment.

NOTE: This chapter is built on the assumption that your software can layer scenes. I will be creating dimmer, color, movement, and other sets of scenes individually with the goal of combining them in a live setting later. If you are using a software program that does not allow for layering, you can still create these movements and colors, but they will have to all be combined into a single scene.

Starting with goals

Whenever I sit down to program lighting for an event, I always run through some of the specifics. I will think about the ages and demographics of those in the crowd, the expectations of the client, and how I will incorporate the lights I have in my inventory into that. If you are a beginner, your lighting inventory may be small, and that's OK! You can use what you have in different and unique ways. Before I ever start programming, I think of how I will lay out my lighting physically. Will I mount my lights vertically, diagonally, or horizontally? What about spacing? How will I position multiples of the same light?

One of the most basic aspects of lighting design is that of symmetry. Symmetry is a natural principle that humans are drawn to – it's pleasing to our eyes. If you attend any live theater production, concert, or club, you will see dozens of lights spread across the trussing. These light shows are often composed of a variety of lighting types in groups of 8, 16, 32, or more. Large groups of lights are spread evenly along the ground or trussing, with an equal number of one type of light on each side of center-stage.

Similarly, you should strive for symmetry in your own light show. If you own two or four moving heads, make sure they are evenly spaced across your T-bar or trussing. The same principle applies to your wash lighting and any other fixtures you may be using. There are exceptions to this rule (such as a lone centerpiece effect light), but on the whole, symmetry will give your light show much more impact and "WOW" factor.

Once I have my lights laid out on my truss or stand, I will make sure that they are imported and laid out in the same way in my lighting program. I'll also create my groups in the program so that I can quickly jump between the different sets of lights as I'm programming. You'll notice I am doing a lot of prep work before even starting to program - in my mind, this is key to having an efficient session. **Programming lights is fun, but if it drags on for hours you can burn out.** Having an efficient workflow that I can follow every time I program helps me stay motivated.

Base Scenes

Once I have everything situated and my lighting program open, there are 2 scenes that I always build first no matter what:

Blackout Scene: The first scene is always a blackout scene, where my dimmer and shutter for all of my lights are both in the off/closed position. Unlike a hardware controller, lighting programs usually don't have a built-in blackout button, so creating a scene for blackout is necessary. I use the blackout scene before the event starts and strategically before big drops in songs to build excitement. I usually program this button as a

"Flash" button so that it can be momentarily activated and deactivated when I release its trigger button.

In-Between Scene: I originally didn't have a scene like this until I began to do lighting for bands. This scene is used between songs with a band as a mellow, low-light transition that provides light to the band on stage to move around and get set up. Even if you aren't doing lighting for a band, an in-between scene is crucial. I usually will just do a static white or cool blue wash and that's it (no effect lights or similar). I may have the moving heads pointed at the stage if there will be someone speaking. The in-between scene is great for any announcements that need to be made when you want the focus to shift to the person speaking.

Another application of the in-between scene is at corporate conferences. In this setting, the in-between scene can be used as your "base" lighting look that you utilize when someone is speaking. You can program this scene with your spotlights at center stage with all of your other lights off or maybe with all of washes on a bright white.

Dimmer Curves

After creating these two scenes, I always begin with some dimmer curves. We learned a little about dimmer curves in the scene builder and generator sections, but here's a quick refresher.

Think of a dimmer curve as a movement pattern for your light's dimmer. If I were to physically grab the dimmer slider for a wash light and push it slowly up and down for 10 seconds, that is a dimmer curve. Alternatively, I could hold the slider at 100% brightness for 1 second, snap it down to 0% for 1 second, and then snap back up to 100%. This is another example of a dimmer curve.

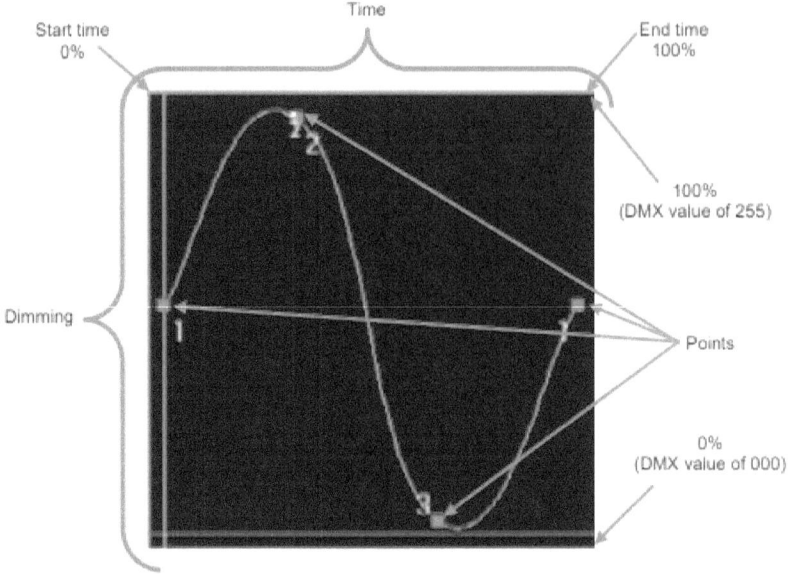

An example of a "Sine Wave" dimmer curve. If I draw this curve on my dimmer channel, my light will fade up smoothly from 50% to 100%, down to 0%, and back to 50%.

Everything we do from this point on will rely on having the lights "switched on", so building some dimmer curves will let us have the lights on while we program the rest of the scenes. Here are a few dimmer curves that I program on a regular basis:

Full-On: This one is pretty obvious. Just throw your dimmer on full and make sure your shutters on your moving heads are open as well. This setting is on most of the time, and it allows me to have live control with my midi controller over brightness. You'll spend a lot of time with your dimmer fully open.

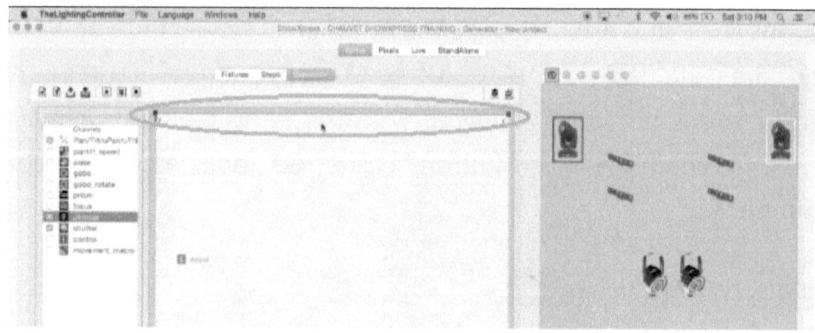

A screenshot of a "full-on" dimmer curve in ShowXpress. I've circled the two points and the line representing the dimmer channel at 100%.

Sine Wave: A sine wave dimmer curve is perfect for adding variety to your lights. Even if you just have one static color on your lighting, having them dim and brighten again can create movement without actually moving anything. Try changing the speed and the phasing (shift) to create more of a chase effect.

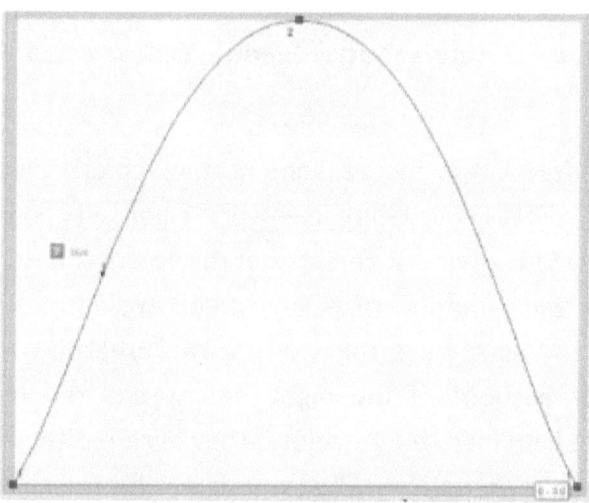

A bell-shaped dimmer curve in ShowXpress, which is just a simplified version of a Sine Wave.

Ramp: The ramp effect causes the dimmer to slowly fade upwards and then suddenly drop to off to 0 (repeating of course). An opposite effect can be achieved with the ramp fading downward and then snapping up to full on. This effect looks especially cool with phasing.

Color Presets

After making the dimmer curves, I make color presets. I normally program colors according to each group of fixtures as opposed to programming a color for all of the fixtures simultaneously. This allows me to change the color of the wash lights for example, while the other lights are another color. If you program a scene with **all** of the lights in your rig the same color, you won't have that flexibility. I'll usually go through and program 8 colors for my wash lights, 8 colors for my moving heads, 8 colors for my effects, etc. Here are the colors I program and the order I program them in:

White – Blue – Purple – Cyan – Green – Yellow – Red – Rainbow Fade

I do this for a specific reason; certain colors elicit certain emotions. I always have white as my first color so I know where to find it easily (it is the most versatile of the colors and fits in almost anywhere). Blue, purple, cyan, and green are all more relaxing, calm, cooler colors. I use them often with breakdowns of songs and slower portions of the night. Yellow and red (along with orange) are very loud, bright colors. They elicit excitement, anger, and other strong emotions. They work great for the higher energy portions of a song. There is a ton to learn about color theory, so feel free to read up on the psychology behind colors to learn more.

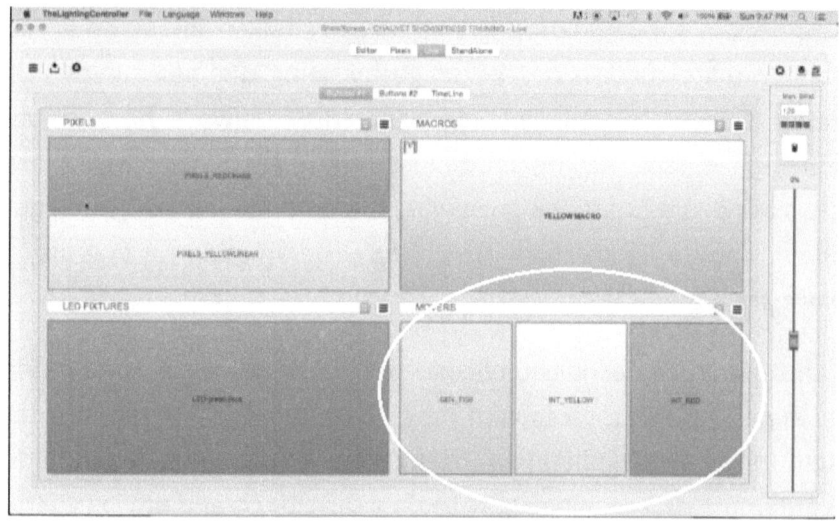

Setting up my boards in ShowXpress with color scenes. I've circled the bottom right board, titled "MOVERS," which contains three color scenes already.

Since I use a MIDI board to trigger my lighting (an Akai APC Mini) I assign these color cues into columns. One column will have my wash colors in it, white at the top to rainbow at the bottom. The next color my moving head colors, and so on. Since they are all in the same order, I just need to know what column controls which light and I can easily find what I need. If you trigger with your keyboard you can find a similar layout.

In addition to static colors, I will also usually program a set of 8-pixel effects. I usually keep pretty close to the same order, creating a white chase effect, a blue fade, a purple chase, and so on. Using pixel (or moving color) effects like this allows you to simulate motion without actually having or using moving lights.

Positions

Having different static positions for your moving heads and

scanners is super helpful. We've already talked about how you can simulate motion on the stage by using dimmer curves and color chases, so your moving lights don't always have to be moving. In fact, having moments where all of your lights are physically stationary helps avoid a bland feeling in your light show. Variety is the spice of lighting programming (or is it life?). Here are a few static positions to consider:

Fan Down/Up: I usually create a "fan" of my lights, both pointing in the air and at the crowd. I have them fan out from the center light (or a pair of lights) toward the sides of the setup. You'll need a minimum of 4 moving lights to do a fan effect, and obviously the more the better.

Cross Down/Up: I'll next create a scene where every other light beam is crossing over each other. Again, what we are doing here is adding variety; the crossing effect is different from the fan we programmed before.

All Right/All Left: For this effect, I will have all of the lights on the left side of my truss point to the back right corner, and all the lights on the right side of my truss point to the left back corner.

Center Spot/DJ Spot: If there is a stage in front of me, I will have a position where all the moving heads and spotlighting a single spot where a speaker would stand. And I usually program a position with all the spotlights on me, cause I'm vain like that (obviously a joke, but it does come in useful at prom when I need to get everyone's attention).

Saw/Zig Zag: With this position, I will have every other fixture either point up in the air or at the crowd, creating two rows of alternating beams. This looks amazing, especially with 6 or more moving heads.

Movements

Movements play a big role in your light show, especially if you are using haze. For those just starting out who don't own moving lights yet, this could mean movement in your effect lighting. If you own moving heads or scanners, this refers to the patterns your lights trace out in the air. If you are in the effect lighting boat, you will most likely be limited to rotational movement. If that's the case, think VARIETY. Slow clockwise rotation, medium clockwise, fast clockwise, counterclockwise, back and forth, etc. If you have moving heads or scanners, here are some of the patterns I create in the ShowXpress generator (or MyDMX effects builder):

Circle: The default pattern when you open a generator or scene builder. The circle is classic and the big movements create amazing looks. Most likely, the large circle shape that by default appears in your generator is too big and your lights will do something that definitely does not look like a circle. This is due to your lights being able to pan greater than 360 degrees and tilt more than 180 degrees. You will have to play with the size and location of your circle in your lighting program to get something that will actually do a circle around your crowd. I usually create two variations of this effect; one with all the heads in sync and one with them out of phase. You can also program this effect at multiple speeds (the same principle applies to all the following shapes).

Figure 8: A figure 8 shape looks like a circle at first but involves more of the beam crossing your crowd as opposed to going around them. With more than 4 moving heads and with the effect out of phase you can get beams and spots crossing all over the crowd.

Sweeping Up/Down & Side/Side: This effect is easy to program;

just have a straight line going through the "generator" box. You will have to find out where to put the line and how long to make it so that you aren't shooting light back behind you (unless you're into that sort of thing). The effect looks great with all the lights in sync, but as with most effects looks AMAZING out of phase.

Zig-Zag: Go ahead and create a zig-zag pattern in your generator and adjust it to keep the light toward your crowd. This effect is similar to the sweeping pattern used before but adds the second dimension (vertical or horizontal) to the mix. Once again, try different amounts of phasing to find what you like best.

Extras

What we have programmed so far is what I would consider the bare necessities for your light show. Combining them will give you hundreds of different combinations, so you're sure to keep your crowd on their feet. After I have taken care of these basics, there are usually a few final things I need to program.

Special Lighting Effects: For some events, I use a Scorpion laser from Chauvet. It doesn't really fit into any of the other categories, so I usually program a few pattern scenes that I like for it and give it a column for itself. This goes for any light you have that doesn't quite fit into the wash/moving head/effect lighting categories.

Strobe Buttons: I program my strobe buttons apart from my main dimmers. I usually program 3 speeds of strobe; slow, fast, and SUPER STROBE (I really call it that). These buttons are always flash buttons, meaning I can momentarily press the button and it will activate the strobe until I release the button. Specific software programs will have instructions for making a flash button. When making these scenes, I activate the shutter (if needed) so that it's open and then adjust the strobe channel until I have the speed I would like. Don't touch the other channels!

Sliders: If you are using a MIDI controller that has sliders or knobs, you can program them into your software. ShowXpress, for example, has master dimmers that can be assigned to MIDI. I will usually assign a master dimmer for the entire rig, and then individual dimmers for each group of fixtures. This keeps me in control of overall brightness and also allows me some manual control over pulsating the brightness to the beat.

Using a MIDI Controller

If you own the full version of MyDMX or ShowXpress, you have the option to use a MIDI controller. MIDI is another electronic protocol that was developed in the 1980s to allow electronic instruments and other digital musical tools to communicate with each other. Many MIDI devices are available such as keyboards, synthesizers, and beat machines. All of them use the same MIDI "language," so you would be technically able to use any of them to control your lighting software.

For lighting purposes, I highly recommend either the **Akai APC Mini** or the **Novation Launchpad** (I promise, I'm not getting paid by either of them). Both can be purchased for around $100 dollars and have dozens of high-quality rubber buttons that can be programmed to your light scenes. These make triggering your light scenes much easier; you're not confined to the cramped keyboard on your laptop and its unusual layout. The Akai APC Mini even has a row of sliders that you can program.

The process of assigning a scene to a MIDI button is easy! Right click on the scene and select "learn MIDI." The computer will wait for a response from your MIDI controller. Just push the button on the controller you want to use for that scene and the computer "learns" it. It's that simple! Before you can go about using your MIDI controller with your program you will need to set it up in the software; don't worry, it only takes a few minutes. Read your software's manual or check out the forums for instruction on this process.

After Programming

After I finish programming all of those scenes into the software and arranging them how I wish, I'm done programming. The process usually takes me under an hour at this point, but took me 2-4 hours when I first began. Have patience with yourself, as learning the quirks and nuances of your specific controller will take time. Don't rush in and try to duplicate what you see at concerts right at the beginning. I did that and burned myself out *really* quick. Pace yourself. Maybe focus on setting up your color palettes and then step away for a drink and a break. Then work on movements.

Take note of how you have your fixtures set up on the trussing. To save you from having to change all of the DMX addresses at your event, you can label the back of each light with a small piece of numbered tape. Just remember which end of your truss you started from! Pack up all of your cabling, your DMX dongle, and your lighting laptop in an organized manner. I keep a separate box for lighting programming so that when I do a lighting gig I can just grab and go. I also keep a hammer for my truss, extra clamps, and various other lighting supplies in the same box.

At Your Event

Playing back your scenes at your event can be handled a number of different ways, depending on your business model. When I have a bride that purchases upgraded dance lighting, I almost always handle the lighting by myself. Even though I will be mixing and MCing at the wedding, I will have the lighting program open on the lighting laptop next to my DJing laptop. Most weddings don't demand you to be changing light scenes every 10 seconds and manually mashing the strobe button constantly, so reaching over to tap a scene change every song or every chorus is easy to manage. **With weddings, I normally will program longer scenes that can be left to run for 2 or 3 minutes without becoming stale.** I still aim to provide as much variety as possible at a wedding without overdoing it. I do keep a strobe button handy at these events to be used SPARINGLY to inject a little energy into a popular song and let people know that the lights are reacting to the music.

When I get hired to specifically do lighting for a dance or a concert, I take a much more hands-on approach. If you ever have watched a lighting designer or operator at a concert it probably seemed as if he was playing his own instrument! Many lighting designers are actively changing and controlling the lights every 32, 16, or even 8 counts! At high energy events like school dances, I will have a trained assistant with me to run the lights (or someone else to DJ). That allows me (or my assistant) to give full attention to the lighting and create active and dynamic light shows. Lighting control at these shows normally involves changing at least one aspect of the lighting (be it color, speed, or brightness) almost every musical phrase. It is important that your timing is precise; a color or position change that falls *exactly* on the downbeat can get those "oohs" and "ahhs" you want.

The key here is **moderation**. Try and spend an equal amount of time in each scene instead of hanging out in your "fast" scenes the whole night. **Make sure that every single light you own isn't**

on the whole night either; shake up the combinations. Try out different color combinations like red and blue, red and white, or just red. Create peaks and valley with the lighting just like the DJ or band does with the music. When the energy is high, bring in those fast chases, movements, and bright colors. When a slow song comes on, freeze the lighting and dim. The possibilities are endless!

Using the tap/beat features of your software

If your lighting program has the ability to use your laptop's microphone for control, this is an additional option available to you. You activate this feature and the scene steps will cycle with the beat of the music as if you were in sound activated mode. ShowXpress allows you to apply this feature to individual scenes, which is important. If applied to the entire show, your moving heads and dimmer curves might behave erratically (they are made up of many steps). This feature is best applied to stepwise color chases that you manually program with the "steps" feature in your software. Another option is the "Tap/BPM" option that will cycle the scene's steps to the BPM you tap out on a key (normally the space bar).

Conclusion

Applying what you've learned

If this book was your first foray into the world of DMX lighting, you've just absorbed a truckload of information. You deserve a drink or something! I hope you have caught the programming bug just like I did back when I first experienced DMX lighting. Take the time to reread this book often, as little details can easily be forgotten as time goes on.

I gained the experience and knowledge that is in this book through years of practice and study, and I hope you never stop striving to learn more. With every year, new lights and technology are introduced into the market that expand the horizons on what is possible with DMX lighting. Spend some time admiring the work of other DJs and lighting designers that you know of, and seek out new inspiration. Watch the hundreds of instructional videos available on YouTube. If you are using MyDMX or ShowXpress, spend an afternoon watching the slew of amazing tutorial videos so that you can understand your software fully.

If you enjoyed this book, I would love to know! Leaving it a 5-star review on Amazon will help it make its way into the hands of other DJs who could use the information. If you have ideas or suggestions for future editions or new books, please leave them in your comments! I wish you all the best in your business and your lighting passion.

-Jordan Nelson

ABOUT THE AUTHOR

Jordan Nelson is a mobile DJ and Master of Ceremonies who owns SLC Mobile DJ in Salt Lake City, UT. Growing up, music was consistently a large part of his life – he began playing piano at age 8, percussion and drum set at 12, and a handful of other instruments throughout high school. He began DJing at the age of 15 and quickly turned this new hobby into a full-fledged business. In the beginning, he cut his teeth performing at school dances and local teen dance parties. With time, his business expanded into the wedding and corporate markets. Today, SLC Mobile DJ is one of the premier DJ companies in Salt Lake City, providing discerning clients with unique and exciting entertainment.

www.ingramcontent.com/pod-product-compliance
Lightning Source LLC
Chambersburg PA
CBHW021445210526
45463CB00002B/639